Geospatial Analyses of Earth Observation (EO) data

Edited by Antonio Pepe and Qing Zhao

Published in London, United Kingdom

IntechOpen

Supporting open minds since 2005

Geospatial Analyses of Earth Observation (EO) data
http://dx.doi.org/10.5772/intechopen.74888
Edited by Antonio Pepe and Qing Zhao

Contributors
Gregory McCarty, Xia Li, Maureen Kapute Mzuza, Weiguo Zhang, Fanuel Kapute, Xiaodao Wei, Christopher Kotsakis, Emiliano Teran, Hartmut Müller, Viviane Després, Daniel Pickersgill, Gilles Grandjean, Xavier Briotet, Karine Adeline, Anne Bourguignon, Audrey Hohmann, Claudia Giardino, Kerttu-Liis Kõks, Rossano Bolpagni, Giulia Luciani, Gabriele Candiani, Moritz K. Lehmann, Mariano Bresciani, Hendrik Jan Van Der Woerd

First published in London, United Kingdom, 2019 by IntechOpen
IntechOpen is the global imprint of INTECHOPEN LIMITED, registered in England and Wales, registration number: 11086078, 7th floor, 10 Lower Thames Street, London, EC3R 6AF, United Kingdom
Printed in Croatia

British Library Cataloguing-in-Publication Data
A catalogue record for this book is available from the British Library

Additional hard and PDF copies can be obtained from orders@intechopen.com

Geospatial Analyses of Earth Observation (EO) data
Edited by Antonio Pepe and Qing Zhao
p. cm.
Print ISBN 978-1-78984-584-6
Online ISBN 978-1-78984-585-3
eBook (PDF) ISBN 978-1-83962-934-1

For EU product safety concerns:
IN TECH d.o.o., Prolaz Marije Krucifikse Kozulić 3, 51000 Rijeka, Croatia, info@intechopen.com or visit our website at intechopen.com.

Meet the editors

Qing Zhao obtained a bachelor's degree in Surveying and Mapping from Central South University, China, in 2004, an MSc in Remote Sensing and Cartography from Wuhan University, China, in 2006, and a PhD in Geo-information Science from the Chinese University of Hong Kong, Hong Kong, China, in 2010. Dr. Zhao joined the Key Laboratory of Geographical Information Science, Ministry of Education, East China Normal University, in 2010. Since 2014, she has worked as an associate professor in the College of Geography Science at East China Normal University. In 2015, she served as a course instructor at New York University Shanghai. Her research interests include synthetic aperture radar interferometry, ocean remote sensing, and air quality monitoring and forecasting.

Antonio Pepe obtained an MSc in Electronics Engineering and a PhD in Electronics and Telecommunication Engineering from the University of Napoli Federico II, Napoli, Italy, in 2000 and 2007, respectively. In 2001 he joined the Institute of Electromagnetic Sensing of the Environment (IREA) of the National Research Council (CNR) where he is a permanent senior researcher. He was a Visiting Scientist at the University of Texas, Austin, in 2005, at the Jet Propulsion Laboratory (JPL), Caltech, Pasadena in 2009 and 2019, and at the East China Normal University, Shanghai, in regular periods from 2014 to 2018. Dr. Pepe acts as a reviewer for several peer-reviewed international journals. From 2012 to 2016 he was also an adjunct professor of Signal Theory and Wireless Communication at the Università della Basilicata, Potenza and University of Naples, Federico II, respectively. He was the recipient of the 2014 Best Reviewer mention of the IEEE Geoscience and Remote Sensing Letters and of the Best Reviewer mention of the MDPI Remote Sensing Journal in 2017.
His main research interests include the development of advanced DInSAR algorithms aimed at monitoring surface deformation phenomena induced by subsidence, volcano activities, and earthquakes, with a particular interest toward the phase unwrapping problems. More recently, he has developed research activities for the generation of DInSAR products through the new generation of SAR instruments, for the generation of hybrid ScanSAR-to-stripmap DInSAR analyses, and for the integration of SAR and optical images.

Contents

Preface XI

Chapter 1 1
Application of Topographic Analyses for Mapping Spatial Patterns of Soil Properties
by Xia Li and Gregory W. McCarty

Chapter 2 33
Clay Minerals Mapping from Imaging Spectroscopy
by Gilles Grandjean, Xavier Briottet, Karine Adeline, Anne Bourguignon and Audrey Hohmann

Chapter 3 47
The Impact of Land Use and Land Cover Changes on the Nkula Dam in the Middle Shire River Catchment, Malawi
by Maureen Kapute Mzuza, Weiguo Zhang, Fanuel Kapute and Xiaodao Wei

Chapter 4 67
Advanced Methods for Spatial Analysis of Bioaerosol Long-Range Transport Processes
by Daniel A. Pickersgill, Hartmut Müller and Viviane R. Després

Chapter 5 79
The Color of Water from Space: A Case Study for Italian Lakes from Sentinel-2
by Claudia Giardino, Kerttu-Liis Kõks, Rossano Bolpagni, Giulia Luciani, Gabriele Candiani, Moritz K. Lehmann, Hendrik Jan Van der Woerd and Mariano Bresciani

Chapter 6 93
Bacubirito: An Outstanding Cosmic Sample on Earth
by Emiliano Terán

Chapter 7 105
Spatial Coordinate Transformations with Noisy Data
by Christopher Kotsakis

Preface

This book presents an overview of some geospatial analyses carried out by using Earth Observation (EO) data and terrestrial measurement campaigns. The chapters cover different aspects related to: (1) the monitoring and investigation of lands and subsurface, (2) the study of atmospheric processes, (3) the investigation of water colour as revealed from satellites, (4) the physical characteristics of an ancient meteorite that impacted on Earth, and (5) the study of spatial coordinate transformations to correctly represent geospatial data in georeferenced systems in the presence of noise.

Chapter 1 addresses the problem of detecting the landscape topography of areas on Earth and its potential modifications over the time. Detecting Earth's surface topography is also important for mapping spatial patterns of soil properties. The chapter provides an overview of the impacts of topographic heterogeneity on the spatial variability in soil properties and presents topography-based models commonly used in soil science. A large-scale soil property map is obtained based on topographic information derived from high-resolution remotely sensed data. The authors show how Remote Sensing (RS) technology enables to obtain valuable information, and this is particularly important in areas with limited data accesses or when it is needed to extrapolate findings from representative sites to larger regions.

Chapter 2 discusses methods for mapping subsurface clay minerals, which is an important issue in terms of mechanics and hydrology. As a matter of fact, clayed soils have a direct impact on ground stability as well as constrain infiltration processes during flooding. The main goal of detecting and characterizing clay minerals in soils is to serve urban planning issues and improve risk reduction by predicting impacts of subsidence on houses and infrastructures. Some recent results are shown in Chapter 2 to characterize clay species and their abundances from spectrometry, used either from a ground spectrometer or from hyper-spectral cameras. The most performing methodology is identified after adequate pre-processing algorithms, coupled with un-mixing methods evaluated from laboratory and real case measurements.

Chapter 3 shows the land use and land cover (LULC) changes over a 26-year period in the middle Shire River catchment, Malawi, Southern Africa, as assessed by using Geographical Information Systems (GIS) and RS techniques. High rate of deforestation averaging 4.3% per annum was observed. Rapid population growth and increase in gross domestic product (GDP) were identified as the major drivers of deforestation and forest degradation due to clearing of vast fields for agriculture, land expansion for urban settlement, and cutting down of trees for wood fuel energy. Combined techniques such as GIS, RS, and socio-economic factors are used. The authors of this investigation provide a clear indication on the potential extension of the proposed methods to other places on Earth where similar challenges occur.

Chapter 4 provides readers with a study on bio-aerosols in the environment. The dynamics and, therefore, the effects on atmospheric processes and the biosphere are often underestimated, or have not yet been sufficiently investigated. Atmospheric models such as FLEXPART and HYSPLIT enable researchers to simulate the transport of particles in the atmosphere and provide information on where air-parcels come from. The chapter shows two methods for combining results of these models with spatial information (e.g., about the vegetation). The former method shows how spatial CORINE land cover distribution can be analysed within the boundaries of HYSPLIT trajectories. The latter proposes FLEXPART simulations that are used in combination with COSMO rain data and tree maps to generate maps that indicate the potential origin of bio-aerosols for selected periods of time.

Chapter 5 addresses RS techniques of water. In particular, it focuses on the monitoring of lakes, which represent inestimable renewable natural resources that are under significant pressure by human activities. RS by satellite sensors offers a significant opportunity to increase the spatial-temporal coverage of environmental monitoring programs for inland waters. Water colour is a water quality attribute that can be remotely sensed and is independent of the sensor specifications and water type. The authors of this investigation used the Multispectral Imager on two Sentinel-2 satellites to determine the water colour of 170 Italian lakes during two periods in 2017. As a result, they found most of the lakes appeared blue in spring and green-yellow in late summer and, in particular, they confirmed a blue-water status of the largest lakes in the subalpine eco-region. They also suggest that information about the colour of the lakes can significantly contribute to the synoptic assessments of the trophic status of lakes.

Chapter 6 analyses the hidden signs of the Bacubirito meteorite impact in Mexico over the areas of its crater. The Bacubirito meteorite is the largest meteorite found in Mexico, the second largest in the Americas, and the fifth largest in the world. It was found in 1863 by the geologist Gilbert Ellis Bailey in the village of Ranchito. It is an iron meteorite weighing between 20 and 22 tonnes; it measures 4.25 meters long, 2 meters wide, and 1.75 meters high. Although largely forgotten by the scientific community after its excavation in 1902, it remains an incredible artifact that has inspired generations of Mexican scientists. Using the Monte Carlo method, an innovative geometrical model and scanner, the authors calculate the precise dimensions and mass (21 tons) of the Bacubirito meteorite's complex structure and provid additional insights into the measurement campaigns and the importance of the meteorite for the cultural heritage of Mexico.

Chapter 7 provides some theoretical insights on the geospatial coordinate transformation to correctly represent geodetic data in noisy environments. The chapter explains the drawbacks of the commonly used approaches and then presents an alternative scheme for spatial coordinate transformations that improves the classic stepwise solution when using noisy coordinates of known stochastic structure. The proposed methodology relies on the joint least squares adjustment of the observed coordinates using their full stochastic model over all points of interest. The mathematical framework and the related properties of this "stacking" approach are presented in detail, along with a numerical example that demonstrates its feasibility for practical problems in geospatial applications.

This book has been made possible by the great work done by the authors and the professional assistance of Publishing Process Manager, Ms. Sara Bacvarova, during all phases of editing.

Dr. Pepe Antonio
Senior Researcher,
National Council Research of Italy (CNR),
Institute for the Electromagnetic of the Environment (IREA),
Napoli, Italy

Dr. Qing Zhao
Associate Professor,
School of Geographic Sciences,
East China Normal University,
Shanghai, China

Chapter 1

Application of Topographic Analyses for Mapping Spatial Patterns of Soil Properties

Xia Li and Gregory W. McCarty

Abstract

Landscape topography is a key parameter impacting soil properties on the earth surface. Strong topographic controls on soil morphological, chemical, and physical properties have been reported. This chapter addressed applications of topographical information for mapping spatial patterns of soil properties in recent years. Objectives of this chapter are to provide an overview of (1) impacts of topographic heterogeneity on the spatial variability in soil properties and (2) commonly used topography-based models in soil science. A case study was provided to demonstrate the feasibility of applying topography-based models developed in field sites to predict soil property over a watershed scale. A large-scale soil property map can be obtained based on topographic information derived from high-resolution remotely sensed data, which would benefit studies in areas with limited data accesses or needed to extrapolate findings from representative sites to larger regions.

Keywords: DTM-based model, high-resolution remotely sensed data, soil carbon, principal component analysis, factor analysis

1. Introduction

Landscape topography is a key parameter influencing biogeochemical processes that occur in the near-surface layer of the earth [1]. In particular, the topography plays an important role in soil formation through regulating soil hydrological regimes and controlling the gravity-driven soil movements [2–6].

Quantitative and qualitative topographic information is essential in understanding the heterogeneity of soil chemistry and physics. Before the 1990s, geographic maps were the main source to quantify landscape topography in soil science [7]. Topographic variables, such as slope and plan and profile curvatures, were calculated manually from these maps to investigate their relationships with soil properties and to generate soil maps [8–10].

Along with the development in computer, aerial, space, and geographic technologies, the availability of high-resolution digital elevation models (DEMs) introduces a new technique in deriving digital terrain models (DTMs) and has been the main source for topographic information extraction in soil biogeochemical studies since the 1990s [7]. A DEM is a digital representation of the terrain surface elevation referenced to a vertical datum. A DTM is an enhanced DEM that has been augmented with breaklines and other observations to describe the land surface geometry [1, 11, 12]. The application of DTMs enables effectively reconstruct topographic landscape over a large scale. Recently, there are two main applications of DTMs in soil science. One is analyses of topographic influences on soil formation and movement, which would be introduced in Section 2. The other is modeling of relations between soil properties and topography and using the results to predict soil properties, which would be discussed in Section 3.

The objective of this chapter is to provide an overview of how topographic heterogeneity causes the spatial variability in soil properties. This chapter starts with an introduction of DTMs applications, which is then followed by reviews of investigations on topographic impacts on soil formation and movements and modeling of soil morphological, chemical, physical properties based on DTMs. The last section presents a case study of DTM-based analysis on how land topography affects soil carbon (C) dynamics.

2. Impacts of topography on soil properties

DTMs are functions of morphometric variables that digitally represent the geometry of the land surface. Various techniques have been developed to generate different DTMs such as topographic metrics of slope, aspect, and curvature. Fifteen topographic metrics that have been reported to be highly correlated with soil properties, including slope gradient, slope aspect, profile curvature, plan curvature, general curvature, flow accumulation, topographic relief, topographic openness, upslope slope, flow path length, downslope index, catchment area, topographic wetness index, stream power index, and slope length factor, are introduced in this section (**Table 1**). Based on the spatial scope, the topographic metrics can be grouped into three main categories [7]:

1. Local topographic metrics: variables describe the surface geometry at a given point on the land surface. Slope gradient, slope aspect, and curvature related (plan, general and profile curvatures) metrics belong to this category.

2. Nonlocal topographic metrics: variables consider relative positions of a selected point, including catchment area, upslope slope, downslope index, flow path length, flow accumulation, topographic relief, and topographic openness.

3. Combined topographic metrics: variables integrate local and nonlocal topographic metrics considering both local surface geometry and relative positions of a point on the land surface. This group of metrics includes topographic wetness index, stream power index, and slope length factor.

These nonlocal and combined topographic metrics often reflect important physics involved in water and soil mass transfer processes considered to have important impacts on soil property patterns.

2.1 Local topographic metrics

Slope gradient indicates the steepness of a line which directly influences the velocity of a gravity-driven flow [2]. For example, a steep area drains quickly and retains less soil than a flat area [4, 13]. Therefore, negative soil redistribution rates with high erosion possibilities are often observed in steep areas. The erosion processes tend to remove fine particles which are usually enriched in soil organic carbon (SOC), leading to low SOC content in a steep area [4, 14]. Meanwhile, the

Variables	Definition and formula
Slope gradient, G (radian)	An angular measure of the relation between a tangent plane and a horizontal plane $G = \arctan\sqrt{p^2 + q^2}$
Slope aspect, A (radian)	Direction of slope measured clockwise with north as 0 $A = \arctan\left(\frac{q}{p}\right)$
Profile curvature, P_Cur (1/m)	Slope change rates in the vertical plane $P\_Cur = -\frac{p^2 r + 2pqs + p^2 t}{(p^2+q^2)\sqrt{1+p^2+q^2}}$
Plan curvature, Pl_Cur (1/m)	Curvature in a horizontal plane $Pl\_Cur = -\frac{q^2 r - 2pqs + q^2 t}{(p^2+q^2)\sqrt{1+p^2+q^2}}$
General curvature, G_Cur (1/m)	Curvature of the surface itself $G\_Cur = -2(r + t)$
Catchment area, CA (m^2)	Upslope area contributing runoff to a given point on the land surface
Upslope slope, Upsl (radian)	Mean slope of upslope area
Downslope index, DI (radian)	Head differences along a flow path $DI = \arctan\left(^{h}/_{L_d}\right)$
Flow path length, FPL (m)	Maximum distance of water flow to a location in the catchment
Flow accumulation, FA (m^2)	Land area that contributes surface water to an area in which water accumulates
Topographic relief, TR (m)	Elevation difference between the highest (h_{max}) point in an area and a given point (h_i) $TR = h_{max} - h_i$
Positive topographic openness, PTO (radian)	Angular measure describing the relationship between surface relief and horizontal distance
Topographic wetness index, TWI	Frequencies and duration of saturated conditions $TWI = \ln\left(^{CA}/_{G}\right)$
Stream power index, SPI	Erosive power of overland flow $SPI = CA_s(Pl\_Cur)\tan(G)$
Slope length factor, LS	Distance from flow origin to a point where deposition begins $LS = (n+1)\left(\frac{CA_s}{22.13}\right)^{0.4}\left(\frac{\sin G}{0.0896}\right)^{1.3}$

L_d *is the horizontal distance to a point with an* h *meter elevation below the starting cell.* CA_s *is specific catchment area.* p, q, r, t, *and* s *are partial derivatives of elevation* (h), $h = f(x, y)$:
$p = \frac{\partial h}{\delta x}$ $q = \frac{\delta h}{\delta y}$ $r = \frac{\delta^2 h}{\delta x^2}$ $t = \frac{\delta^2 h}{\delta y^2}$ $s = \frac{\delta^2 h}{\delta x \delta y}$.

Table 1.
Definitions and formulas of selected topographic metrics.

slope gradient can impact soil water content [2, 7]. For relatively flat areas, soil water content commonly decreases with slope gradient due to increased lateral flow and depositional crusts that decrease infiltration; while as slope steepens, rills may occur that can disrupt the crusts and favor greater infiltration, and therefore lead to a positive relationship between soil water content and slope gradient [15–18].

Slope aspect shows the direction that a slope faces. This metric influences soil-water balance by affecting insolation and evapotranspiration [19]. Soil temperature and evapotranspiration tend to be lower, and soil water content tends to be higher in shady aspect areas. These environmental conditions can be favorable for slow decompositions of organic matter and high accumulations of soil C and nitrogen (N) content [20–22]. Soil water content impacted by slope aspect can further influence vegetation density, which may have impacts on runoff velocities and soil erosion rates [23].

Profile, plan, and general curvatures are important topographic factors controlling patterns of overland flow and soil water content. Profile curvature shows upwardly concave with positive values and upwardly convex with negative values (**Figure 1a**). This variable affects flow acceleration and deceleration and therefore influences soil redistribution and distribution patterns of SOC content [24–26]. A positive plan curvature value indicates a laterally convex surface and a negative value indicates a laterally concave surface [24] (**Figure 1b**). Water accumulates and soil water content decrease when flow diverges (positive plan curvature) and increase when flow converges (negative plan curvature) [7]. General curvature is the curvature of the land surface and describes peaks with positive values and valleys with negative values. This metric enables more accurate estimation of overland flow paths than plan and profile curvatures, and can significantly correlate with patterns of soil erosion and deposition [4, 27].

2.2 Nonlocal topographic metrics

Catchment area and slope related nonlocal topographic metrics (upslope slope and downslope index) affect soil properties mainly through regulating soil water content. At a location, increases in water amount from upslope areas can increase water supply to the location and affect the water accumulation [28]. Therefore, positive correlations have been observed between the catchment area and soil water content [7]. Furthermore, as the catchment area increases, the chance for sediment deposition increases, and thus affects the soil C stocks [29]. The upslope slope relates to slopes in upslope contributing areas. Overland flow velocities are usually less at positions with lower values of upslope slope [1, 30]. The downslope index is a metric including dispersal (downslope) controls on drainage [31]. Since the drainage of a location is the balance between the water from a specific upslope contributing area and to a downslope area, this index usually shows a better representation of groundwater gradients and soil water content than slope gradient [31, 32].

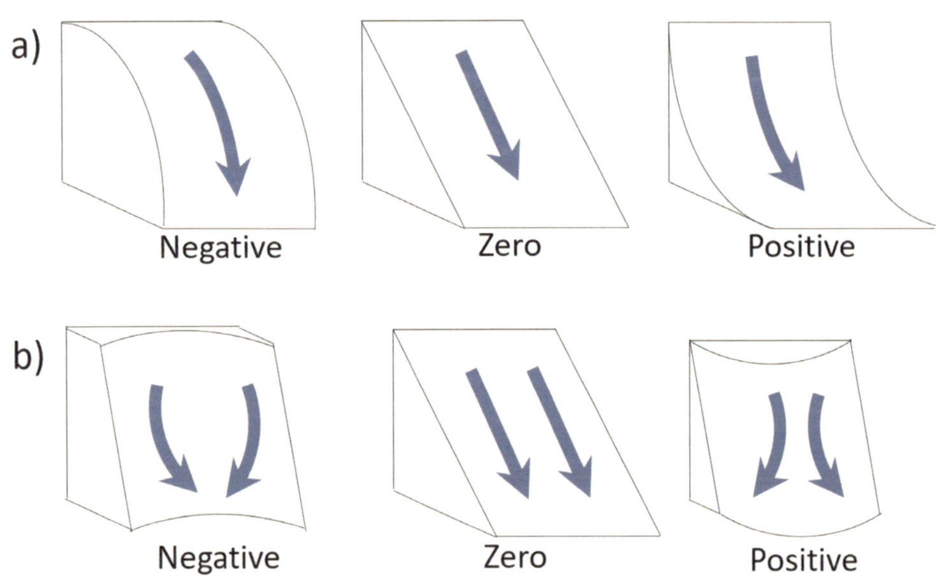

Figure 1.
Diagrammatic illustration of (a) profile and (b) plan curvatures.

The two flow-related nonlocal topographic metrics (flow path length and flow accumulation) reflect the impacts of soil hydrology on soil properties. The longer flow path length decreases overland flow velocity and increases infiltration [33, 34]. Increased erosion of fine particles can also be observed when flow path length increases [35, 36]. This metric has been widely used in soil erosion models, describing soil loss under flow divergence and convergence conditions [37, 38]. Flow accumulation mainly influences water conditions in soils. Flow volume and soil water content response positively to this metric, which in turn can influence the soil C stocks [39, 40].

For topographic relief, higher values suggest larger differences from the highest points, which would stimulate flow velocity, leading to more rapid downslope soil transport from low relief areas [4, 41]. Moreover, topographic relief influences landscape drainage characteristics. Tucker and Bras [42] found that drainage density was positively correlated with relief in semiarid and low-relief landscapes but negative related to relief in humid or high-relief landscapes. Areas with a broad range of relief may cover several altitudinal climatic zones with differences in vegetation types, further influencing weathering and denudation processes [43].

Topographic openness describes the distinction between relief and surrounding topographic features [44]. Convex landforms often exhibit high positive topographic openness values, whereas concave landforms typically have high negative topographic openness values (**Figure 2**) [44, 45]. Therefore, soil water content may change with this variable [46]. The low positive openness areas are more likely to be depressional areas with high soil water contents that provide suitable anaerobic environments for denitrification but impede aerobic SOC decomposition [4, 46].

2.3 Combined topographic metrics

Topographic wetness index combines a local topographic metric (slope) and a nonlocal topographic metric (upslope contributing area) [47]. It is considered as an indicator effectively reflecting the spatial distribution of wetness conditions as the upslope contributing area would impact groundwater level and soil water content, and the slope would influence drainage processes [3]. Areas with higher wetness index tend to be wetter. The topographic wetness index has been used to estimate the spatial distribution of hydrological and geochemical properties of soil, and significant correlations have been observed between this metric and soil C and N content [3, 4, 32, 46, 48, 49].

Stream power index takes into account both specific contributing area and slope. This metric is useful for characterizing potential erosive power of water flow [1]. When the slope gradient and catchment area increase, the amount of water from contributing area and the velocity of water flow increase, and consequently

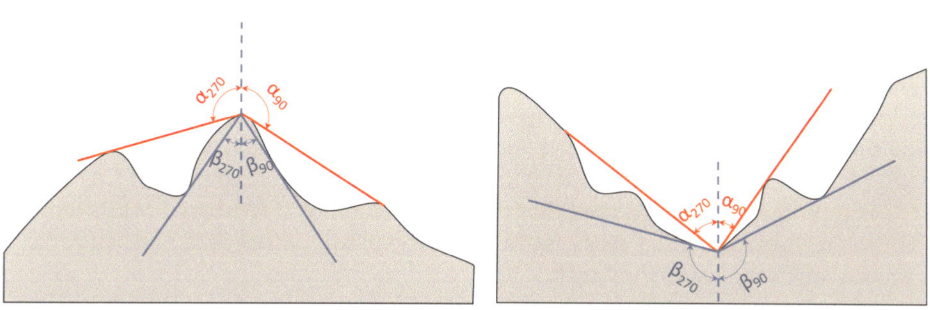

Figure 2.
Positive (α) and negative (β) topographic openness along two profiles.

enhancing the erosive power of water [50]. Therefore, areas with larger stream power index values have greater potential to be erosive regions [4, 51]. Due to its impacts on erosive power, this metric can also be useful in understanding erosion-induced soil C and N dynamics [14, 52, 53].

Slope length factor includes the length and steepness of a slope and thus reflects the topographic impacts on erosion [54, 55]. As the slope length increases, the soil loss per unit area usually increases due to a greater runoff accumulation on a longer slope length that increases transport capacity of runoff; as slope steepness increases, soil loss also generally increases [54, 56]. This factor is essential in estimating soil transport and erosion by runoff [37, 38, 50, 56, 57].

3. DTM-based soil property prediction

In a DTM-based soil property model, the predictive variable could be the morphological, chemical, or physical property of soil. Development of DTM-based models follows two assumptions including that (1) the controls of topography on soil properties can be found through a relatively small set of soil samplings and topographic metrics, and (2) the statistical correlations between topography and soil properties are often strong. In this case, soil properties can be predicted based on the topographic metrics [7]. Due to the recent availability of large-scale, high-resolution DEMs, DTMs over large-scales can be derived. The DTM-based models benefit investigations in regions with limited observations and can generate spatially continuous soil property maps based on extrapolation.

Methods of DTM-based soil property prediction could be grouped into two categories [7]:

1. DTM-based models to predict quantitative soil properties based on statistical analyses. Multiple regression analysis, regression kriging, cokriging, and kriging with external drift are the widely used methods to predict quantitative soil properties.

2. DTM-based models to predict categorical soil properties. Statistical methods such as classification tree model, fuzzy logic, and discriminant analysis are usually employed in this category.

3.1 DTM-based methods to predict quantitative soil properties

Multiple linear regression (MLR) simulates relationships between two or more independent variables and a dependent soil property variable by fitting to a linear equation. The DTM-based MLR models have been applied to study spatial patterns of soil structures, horizonation, and soil water content [12, 39, 58–65], to explore spatial variability of cation exchange capacity and pH [62, 65–67], and to derive continuous quantitative maps of SOC, C isotopes, and nutrients over large spatial scales [4, 6, 12, 14, 46, 52, 68, 69]. In some modeling investigations, Hybrid regression methods were used to improve the efficiency of soil property prediction. Li et al. [4] combined stepwise MLRs with principal component analysis (PCA) for SOC mapping. Results suggested that the combination of DTM-based MLRs with PCA outperformed regular stepwise MLRs in the prediction of SOC and soil redistribution rates at a watershed scale.

Regression kriging (RK) is a spatial prediction combining an MLR with kriging of the regression residuals. The RK acts as a MLR model if the data used in the model have low spatial structure, and the method reduces to Ordinary kriging (OK) if there are no linear statistic correlations between the dependent variable and the ancillary variables [64]. Based on topographic and other environmental variables, numerous studies have applied RK for predicting spatial patterns of soil properties, such as soil horizon thickness [70, 71], soil structures [63, 70, 71], soil water content [72, 73], soil C content [63, 69, 74, 75], cation exchange capacity [66, 76], and soil hydraulic properties [77, 78]. Generally, this method is more accurate in soil property estimations than the OK, Cokriging, or MLR because residual values from kriging analysis were added to the regression [63, 70, 71, 73, 76, 77]. However, Zhu and Lin [72] reported that the RK performed worse than the OK for soil property prediction in relatively low relief areas.

Cokriging (CK) and kriging with external drift (KED) are also popular and practical spatial predicting techniques in digital soil mapping. The CK calculates soil properties by investigating topographic metrics in the kriging procedure and KED uses external ancillary topographic variables as kriging weights. Various studies have employed CK and KED to derive continuous maps of soil physical and chemical properties [65, 66, 70, 71, 76–80]. Some of these studies were also suggested that these techniques would be superior to OK in soil property estimation when the selected topographic metrics are highly correlated with the dependent variables [79, 81].

3.2 DTM-based methods to predict categorical soil properties

Classification tree models (CTMs) are a major type of Decision Tree method used in soil science, in which the target variable is a categorical soil variable. This model applies a set of rules that use explanatory variables to split data into homogeneous subsets. The explanatory variables can be either categorical, such as geological unit number, soil unit, etc., or continuous, such as slope, elevation, topographic wetness index, etc., [82, 83]. Compared to mathematic functions, the tree structure can provide a more visualized explanation of relationships between explanatory variables and the target variable. The CTMs can be used to derive efficient predictions of soil taxonomic classes from local to large spatial scales [82–91]. Soil drainage can also be effectively classified using the CTMs with soil profiles and topographic metrics as predictors [92–96].

The basic idea of fuzzy logic (FL) is to show "degrees of truth" for a variable. Soils are continuums in both geographic and attribute spaces. As a result, using 0 and 1 or discrete categories cannot provide sufficient information about soil properties. The FL overcomes the limitation. If a variable belongs to a set, the model would take a value between 0 to 1 instead of 0 or 1. Several studies have used the DTM-based FL to improve soil taxonomic classes in soil mapping [97–101], soil texture and soil horizonation prediction [98, 102–105], and soil vulnerability classification [106]. Qi et al. [102] found that using the FL the accuracy of soil series name prediction increased 17% compared to the conventional soil survey. The FL was also combined with maximum likelihood regression to derive the prediction of some continuous soil properties [97].

Discriminant analysis (DA) is a type of supervised classification to assign objects to the most likely group among a lot of groups. It uses some observations (training dataset) to classify others. This method is applicable when correlations between soil property variables and independent variables are high [107]. It has been applied to differentiate soil taxonomic classes and to generate soil texture maps using multiple ancillary variables including topographic metrics [107–110]. Several studies also demonstrated the feasibility of using DA in deriving soil drainage classes based on its relationships to topography and soil electrical conductivity [111–114].

3.3 New emerging methods to predict soil properties

Artificial intelligence (AI) or machine learning gives computers a degree of sophistication to act intelligently [115]. Therefore, to be intelligence, computers should be able to learn from training datasets, correctly interpret external data, and apply learned knowledge to achieve specific goals. With increased computing power, massive sets of labeled data, and developed pre-trained models, increasingly researchers have applied AI to fields such as speech recognition, objective detection, visualization, machine translation, image processing and others [116]. However, it is not until the recent decade that the potential applications of AI on soil property prediction have come into more common awareness by scientists.

Artificial neural networks (ANNs) are a representative AI technique that has been applied to solve complex machine learning problems (**Figure 3a**). The method has similar data processes as a biological neural with nonlinear mapping structures, which consists of a set of interconnected units (neurons) [117]. The input neurons are predictors, linking to one layer of hidden neurons and finally linking to the output variables [118]. To obtain accurate prediction results, the network model is trained first by a set of observations. The weights that connect neurons are adjusted iteratively using the training dataset. After training, the model is applied to predict areas with the same input variables. ANNs outperform traditional statistics in handling large datasets even when the input data are noisy with low levels of precision due to the ability to reduce bias by evenly distributing training data across classes [119]. Various researchers have employed ANNs for efficient prediction of quantitative soil chemical and hydrological properties [118, 120, 121] and adequate mapping categorical soil taxonomic classes [122–129] based on DTMs and environmental variables. Zhao et al. [93] also tested the feasibility of using ANNs for soil drainage classification and found an accuracy of 52% between field observations and digital classification.

Deep learning (DL) is considered as an advanced ANN (**Figure 3b**) that has been facilitated by recent advances in technology for highly parallel computing. In contrast to single hidden layer ANNs, DL algorithms allow the computer to learn on its own by multi-layer nonlinear transformations of the input training data [130]. For instance, such algorithms can define edges within images by training on multiple examples and perform automatic feature extraction without human intervention. Therefore, massive quantities of representative learning data are the prerequisite for effective estimation from DL. The architectures of DL include Convolutional Neural Networks (CNNs), Deep Belief Networks (DBNs), and

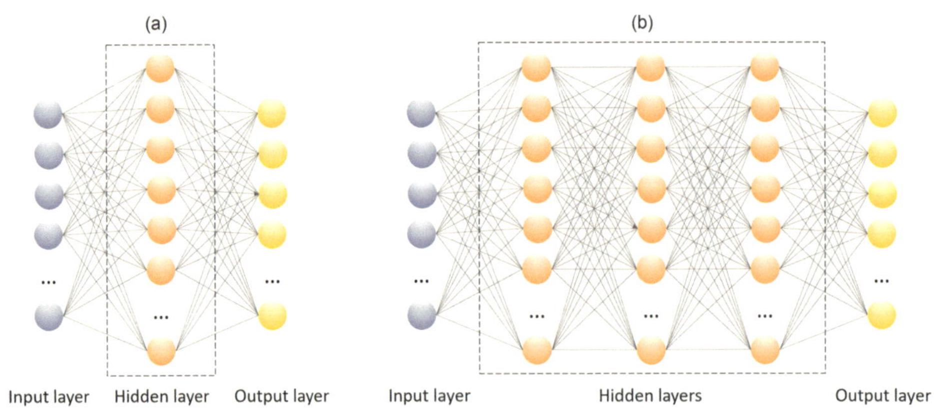

Figure 3.
Architectures of the (a) artificial neural networks and (b) deep learning.

Recurrent Neural Networks (RNNs). CNNs are classic feedforward networks in which the hidden layers consist of convolutional layers, pooling layers, and fully-connected layers. The convolutional layers apply different convolution operations (filters) to pass results from local patches in the feature maps of the input or the previous layers to the next layers, enhancing certain features in the output. Neurons in the same convolutional layer share the same weight. The pool layers merge similar feature together, improving the robustness of features against noise and distortion. The convolutional and pooling layers are finally stacked to a fully-connected layer. The local connectivity in a convolution layer allows CNNs to achieve a better generation in output analysis and the shared weights increase the possibility to extract information of high complexity [131]. Padarian et al. [132] applied CNNs to predict SOC at multiple depths using elevation, slope, topographic wetness index, temperature and rainfall as input data. The results suggested that the CNNs reduced errors by 25% for SOC predictions than the conventional Cubist model. The DBNs are considered as a composition of unsupervised sub-networks, which are trained to maximize the likelihood of training data. Each sub-network serves as a visible layer used for unsupervised training of the next layer [133]. Song et al. [134] demonstrated the usefulness of DBNs in predicting soil water content in highly nonlinear forms over an irrigated field. RNNs have a "memory" called hidden state to remember all information that has been calculated, so the output of RNN loops connect to their past decision nodes based on the hidden state. The networks process an input sequence at one time, preserving the sequential information in the hidden state and producing the output sequence. Therefore, this model is especially useful for tasks containing sequential input [131]. Researchers have demonstrated the feasibility of using RNNs for hydrological study, although no reports were found using this approach to map soil properties based on topography [135, 136].

Random forest (RF) is another emerging method of AI and consists of an ensemble of classification and regression trees for prediction (**Figure 4**). Each tree is a random subset of features and uses a random set of the training data (about 2/3 of the available observations), which increases the diversity of the forest and decreases the correlation of individual trees. RF commonly has high efficiency and

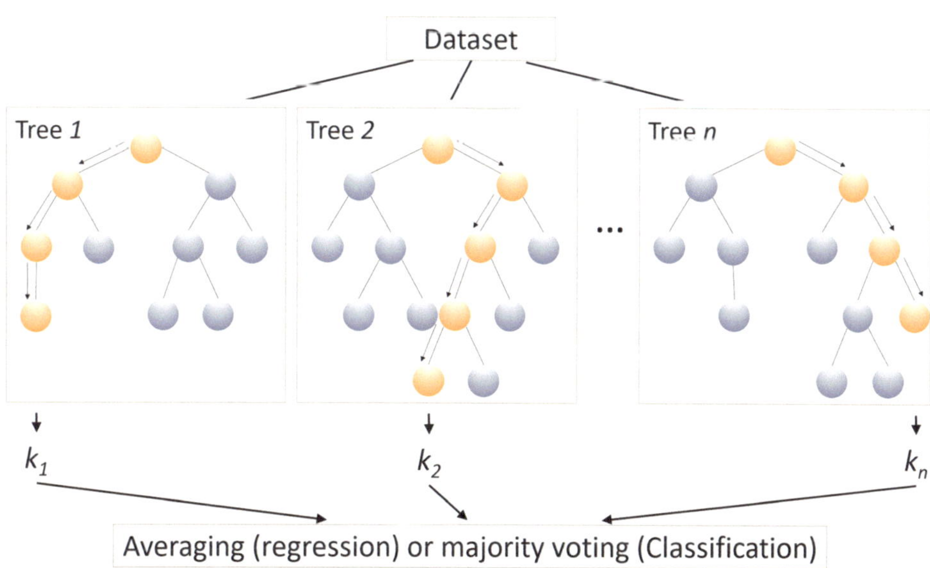

Figure 4.
Architectures of the random forest model.

low bias and variance since the output is the average or majority voting of a large number of trees [137]. The method has been proved to be resistance to over-fitting because each tree is trained on a unique bootstrap subset and provide a reliable error estimate using Out-Of-Bag data (the remaining one-third of the observations) [138]. Because of the above advantages, increasing scientists have used RF in soil mapping. Combining topographic metrics, environmental variables, climate variables, or/and land cover as input, RF can predict quantitative and category soil properties. For quantitative soil property prediction, the output is the average of individual tree outputs. RF has been successfully applied to investigate spatial patterns of soil organic matters [139–142] and to estimate soil texture [143]. Guo et al. [144] further developed soil organic matter prediction by combining RF with Residuals Kriging, for which the prediction accuracy increased dramatically (R^2 = 0.86) compared to the method using RF only (R^2 = 0.65). For categorical soil property classification, the output is obtained from voting by the majority on the correct classification. Several studies have demonstrated the feasibility of using RF for updating soil survey maps [145] and predicting soil classes in unmapped regions [146–148].

4. Case study: DTM-based models on SOC dynamics

4.1 Introduction

In this section, a case study about DTM-based modeling of SOC and soil redistribution (SR) was discussed to understand the impacts of topography on SR and SOC dynamics. We also compared efficiencies of three types of DTM-based models in predicting the soil properties. Cesium-137 (^{137}Cs) was used to trace the SR process, and high-resolution light detection and ranging (LiDAR) data were applied to derive DEMs for DTM extraction. Based on the DEM-derived topographic information and field measured SOC density and SR rates, the multiple linear regression (MLR), MLR combined with principal component analysis (MLR-PCA), and MLR combined with factor analysis (MLR-FAn) were developed and discussed.

The study was carried out in Walnut Creek watershed (WCW), which is located in Boone and Story counties, Iowa, USA (**Figure 5a**, 41°55′–42°00′N; 93°32′–93° 45′W). It has a humid continental climate. The landscape of this watershed is relatively flat with a low topography relief (2.03 ± 1.62 m). The typical soils are poor-drained Nicollet and Webster soils in the lowlands and well-drained Clarion in the uplands. More than 86% area of the watershed is cropland. Chisel plowing in autumn and spring disking are the current primary tillage operations. Directions of tillage practices in the WCW are mostly north-south or east-west, depending on the management and field configurations. Detailed information on climate, soils, and farming practices can be found in Hatfield et al. [149].

Two field sites were selected for intensive sampling investigation. Each site is approximately 15 ha. Site 1 is in the WCW (**Figure 5b**) and Site 2 is located between Boone and Ames (**Figure 5c**), which is within 10 km of the closest watershed boundary. Similar to the WCW, low reliefs (<4.6 m) were observed for both sites. Tillage practices at these two sites were along the north-south direction.

4.2 Materials and methods

4.2.1 Field sampling and laboratory analyses

The SOC and ^{137}Cs data used in this section have been reported in Ritchie et al. [25] and Li et al. [4, 52]. A total of 460 locations were randomly selected for WCW

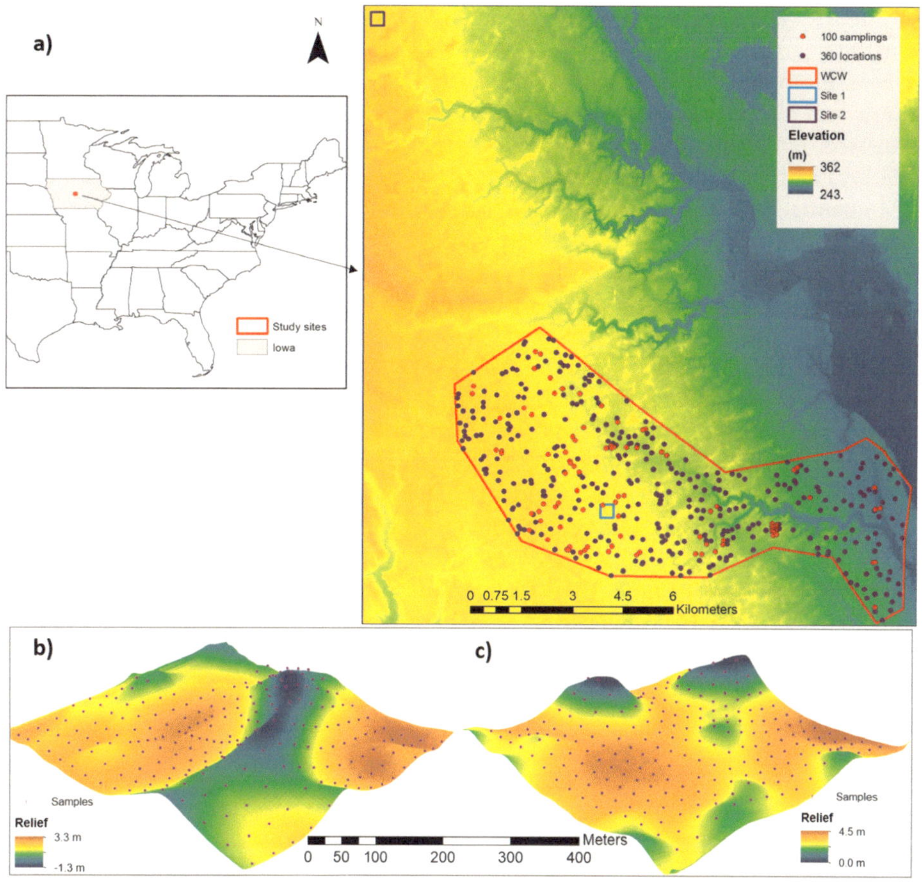

Figure 5.
Location of a) Walnut Creek watershed (WCW), b) Site 1 and c) Site 2 (z-axis 15× elevation).

and 230 locations were selected for each site of Sites 1 and 2 (**Figure 5**). Topography information was extracted for all locations using the LiDAR-derived DEMs. For the watershed, 100 out of the 460 locations, including two 300-m transects, were chosen for field estimations of SOC content and ^{137}Cs inventory in 2006. The field samplings for Sites 1 and 2 were collected in 2003. A 25 × 25 m grid was created for each site of Sites 1 and 2. The 230 samplings were obtained at grid nodes. At each location, we collected three samples that were located within a 1 m × 1 m quadrat from top 30 cm of soil using a push probe (3.2 cm diameter). At locations where sediment depositions were observed, deeper soils from the 30 to 50 cm layer were collected. Four reference soil samples for estimation of the baseline ^{137}Cs inventory were collected from a local cemetery in WCW where no apparent soil redistribution had occurred from the 1950s. Trimble RTK 4700 global positioning system (GPS) was used to record the locations of sampling.

During laboratory analyses, bulk density was calculated after drying soil at 90°C for 48 hours based on the soil volume and dry mass weight. Then, the three samplings were mixed and sieved through a 2 mm screen. We ground subsamples that were taken from the composite soils to fine power with a roller mill and measured soil total C content by dry combustion at a temperature of 1350°C using an elemental analyzer (LECO CNS 2000, LECO Crop., St. Joseph, MI). Then, C content in $CaCO_3$ was analyzed by dry combustion after the soil sample was baked in a furnace at 420°C for 16 hours. Estimates of SOC content ($SOC_{content}$, %) were obtained from the differences between total C content and C content in $CaCO_3$.

SOC density (ρ_{SOC}, kg m^{-2}) of the top 30 cm layer was calculated from the bulk density (ρ_{bulk}) and SOC content using the equation of $\rho_{SOC} = SOC_{content} \times \rho_{bulk} \times 0.3$.

Measurement of ^{137}Cs inventory used another subsample of the sieved soil sample and placed and sealed in a Marinelli beaker. The ^{137}Cs concentration was estimated by Canberra Genie-2000 Spectroscopy System that receives input from Canberra high purity coaxial germanium crystals (HpC > 30% efficiency) into three 8192-channel analyzers through gamma-ray analysis. Analytic mixed radionuclide standard (10 nuclides) that follows the U.S. National Institute of Standards and Technology was applied for calibration the Spectroscopy System. The measurement precision is between ± 4 and $\pm$ 6%. Unit of ^{137}Cs concentration is in Becquerels per gram (Bq g^{-1}) and was converted to ^{137}Cs inventory in Becquerels per square meter (Bq m^{-2}) using soil bulk density.

Calculation of SR rates based on ^{137}Cs inventories was carried out by applying a Mass Balance Model II in a spreadsheet Add-in program [150]. Before running the model, parameters of tillage depth, proportion factor, and relaxation depth were set to 0.25 m, 0.5, and 4 kg m^{-2}, respectively. The baseline ^{137}Cs inventory estimated from the mean of ^{137}Cs inventory in Ref. sites was 2657 Bq m^{-2} for Sites 1 and 2 in 2003 and 2526 Bq m^{-2} for the WCW in 2006. Positive SR rates were obtained when ^{137}Cs inventories were higher than the baseline and the sites were referred as depositional sites; while eroded sites were considered when negative soil SR rates were estimated under conditions of lower ^{137}Cs inventories than the baseline. Details of soil sampling and laboratory analyses can be found in Ritchie et al. [25] and Li et al. [4].

4.2.2 Topographical analysis

Fifteen topographic metrics that were discussed previously were used as ancillary variables for the development of the DTM-based SOC and SR models. All metrics were derived from DEMs generated from high resolution (1 m horizontal and 0.1 m vertical resolutions) LiDAR data [48]. Before generation of topographic metrics, inverse distance weighted interpolation was applied to produce 3 m spatial resolution DEMs after converting the raw LiDAR data to LAS files.

Topographic metrics were derived based on the 3 m DEMs after filtering twice by a 3-kernel low pass filter. Modules in an open-access software of the System for Automated Geoscientific Analysis (SAGA) v. 2.2.5 were applied to generated 14 of the selected topographic metrics including slope gradient (G), aspect (A), profile curvature (P_Cur), plan curvature (Pl_Cur), general curvature (G_Cur), flow accumulation (FA), positive topographic openness (PTO), upslope slope (Upsl), flow path length (FPL), downslope index (DI), catchment area (CA), topographic wetness index (TWI), stream power index (SPI), and slope length factor (LS). Topographic relief (TR) was calculated by the difference between a maximum elevation map within a specific area and the filtered 3 m DEMs. In order to reduce errors due to an arbitrary selection of the radius of the specific area, a series of maximum elevation maps with multiple radiuses including 7.5, 15, 30, 45, 60, 75, and 90 m, were used to generate TR maps with different spatial scales. Principal component analysis (PCA) and varimax rotated Factor Analysis (FAn) were used and converted the TR maps into two main topographic relief components (TRPC1 and TRPC2) and two topographic factors (TRFA1 and TRFA2). The detailed topographic metric processing can be found in Li et al. [52].

4.2.3 Statistical analysis

Spearman's rank analysis was applied to understand the impacts of topographic metrics on SR and SOC distribution patterns. Due to high correlations between some of the topographic metrics, PCA and varimax rotated FAn were used to limit errors caused by collinearity between topographic variables. The PCA analyzed topographic metrics from the 460 locations of the WCW. Loadings for the first eight components that explained 90% of the variance of all metrics were selected and used to calculate topographic principal components (TPCs) at the field Sites 1 and 2. Similarly, eight topographic factors (TFAs) at field sites were also estimated based on loadings from the watershed using FAn with varimax rotation.

4.2.4 Model calibration and evaluation

The stepwise linear regression with "leave-one-out" cross-validation was applied for MLR, MLR-PCA, and MLR-FAn model development using the topographic metrics from two field sites. Akaike Information Criterion was used to select variables contained in each model. The SOC density and SR rates were log-transformed to meet the assumption of residual normality. Model efficiencies were assessed with the following three criteria. The first one is the adjusted coefficient of determination (R^2_{adj}), which adjusts coefficient of determination based on the number of predictors in the model. The second one is Nash-Sutcliffe efficiency (NSE). Ranging from $-\infty$ to 1.0, the NSE estimates the ratios of residual variance to measured variance. The model performance was considered acceptable when the NES is in a positive value. The third one is the ratio of the root mean square error (RMSE) to the standard deviation of measured data (RSR). It standardizes RMSE. The smaller the RSR value is, the higher efficiency it indicates. Usually, the model performance is considered as satisfactory if the NSE value is larger than 0.5 and the RSR is <0.7 [151].

4.3 Results and discussion

4.3.1 Topographic impacts on soil properties

The high-resolution topographic metrics derived from LiDAR data presented detailed topographic information in the WCW. Take field Site 1 as an example, **Figure 6** exhibited characteristics of each topographic metric in response to the elevation. Seven topographic metric maps, including catchment area (CA, **Figure 6f**), downslope index (DI, **Figure 6h**), flow path length (FPL, **Figure 6i**), flow accumulation (FA, **Figure 6j**), topographic relief component 1 (TRPC1, **Figure 6k**), topographic relief factor 1 (TRFA1, **Figure 6m**), and topographic wetness index (TWI, **Figure 6p**), showed high values in depressional areas and low values in sloping and ridge areas. Positive topographic openness (PTO, **Figure 6o**) had a reverse pattern compared to the above seven metrics. It showed high values in ridge areas where a wider view of a landscape can be seen. For slope gradient (G, **Figure 6a**), upslope slope (Upsl, **Figure 6g**), topographic relief component 2 (TRPC2, **Figure 6l**), topographic relief factor 2 (TRFA2, **Figure 6n**), stream power index (SPI, **Figure 6q**), and slope length factor (LS, **Figure 6r**), high values were observed in sloping areas, but low values were found in ridges and depressional areas.

Most topographic metrics showed significant correlations with SOC density and SR rates except A. The A was slightly correlated with SOC density ($r = -0.097$; $P = 0.02$) and insignificantly correlated with ^{137}Cs inventory ($P > 0.05$) and SR rates ($P > 0.05$). Generally, stronger topographic controls on SOC density than ^{137}Cs inventory and SR rates were observed (**Table 2**). TWI, TRFA1, TRPC1, CA, FPL, DI, FA, SPI, and TRFA2 were significantly positively correlated with SOC density and G, LS, PTO, Upsl, Pl_Cur, G_Cur, TRPC2, and P_Cur were significantly negatively correlated with SOC density. For both ^{137}Cs inventory and SR rates, similar

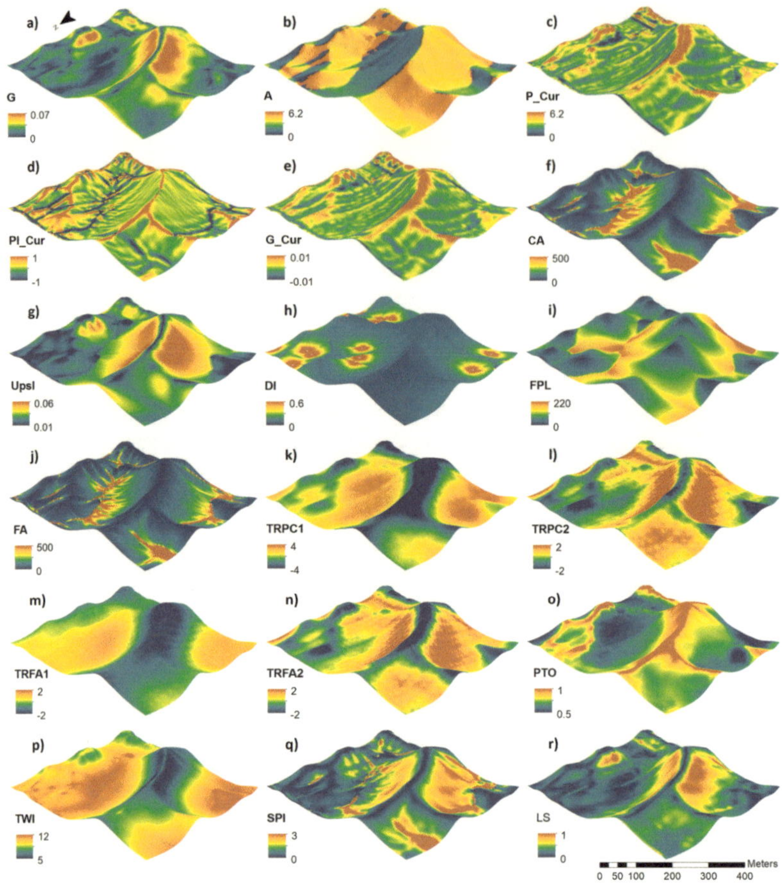

Figure 6.
Topograhic metrics of Site 1 (z-axis 15× elevaton). The metrics include a) slope gradient (G), b) aspect (A), c) profile curvature (P_Cur), d) plan curvature (Pl_Cur), e) general curvature (G_Cur), f) catchment area (CA), g) upslope slope (Upsl), h) downslope index (DI), i) flow path length (FPL), j) flow accumulation (FA), k) topographic relief principal component 1 (TRPC1), l) topographic relief principal component 2 (TRPC2), m) topographic relief factor 1 (TRFA1), n) topographic relief factor 2 (TRFA2), o) positive topographic openness (PTO), p) topographic wetness index (TWI), q) stream power index (SPI), and r) slope length factor (LS).

high related topographic metrics ($|r| > 0.5$) were observed, including TRPC1, TWI, TRFA1, G, and CA.

Topographic wetness index (TWI) was the most influential topographic factor with a correlation coefficient up to 0.735. The finding is consistent with the high TWI impacts on SOC in previous studies. The high impact of TWI suggests that soil water content distribution was an important driver of SOC dynamics in the WCW. In areas with high TWI and possibly elevated water content, litter decomposition rates decrease and plant productions increase, which increases SOC input and accumulations and results in high SOC density in the soil; while low soil water content areas provide an adequate environment for rapid aerobic decomposition of soil C, leading to a negative correlation between SOC and TWI [3, 4, 32, 46, 152].

Topographic relief (TR) was found to be the most important factor for ^{137}Cs inventory and SR rates with correlation coefficients of 0.686 and 0.687, respectively. This metric was also highly correlated with SOC density. The strong effects of TR on soil properties may be due to its influence on flow velocity. The flow velocity reflects runoff shear stress, which would impact the sediment transport capacity of the runoff [153, 154]. Thus, as the TR increases, the flow velocity and

	G	A	P_Cur	Pl_Cur	G_Cur	CA	Upsl	DI	FPL
SOC	−**0.687*****	−0.097*	−0.159**	−0.333***	−0.288***	**0.587*****	−0.315***	0.413***	**0.500*****
^{137}Cs	−**0.646*****	—	−0.210***	−0.277***	−0.286***	**0.568*****	−0.209***	0.366***	0.490***
SR	−**0.650*****	—	−0.205***	−0.274***	−0.282***	**0.565*****	−0.217***	0.361***	0.487***
	FA	TRPC1	TRPC2	TRFA1	TRFA2	PTO	TWI	SPI	LS
SOC	0.204***	**0.698*****	−0.171***	**0.723*****	0.143**	−0.451***	*0.735****	0.165***	−0.453***
^{137}Cs	0.192***	*0.686****	−0 095*	**0.623*****	0.221***	−0.437***	**0.640*****	0.160**	−0.433***
SR	0.193***	*0.687****	−0.099*	**0.624*****	0.218***	−0.427***	**0.647*****	0.156***	−0.438***

G and A are slope gradient and aspect, respectively; P_Cur, Pl_Cur, and G_Cur are profile curvature, plan curvature and general curvature, respectively; TRPC1 and TRPC2 are topographic relief components 1 and 2, respectively; TRFA1 and TRFA2 are topographic relief factors 1 and 2, respectively; PTO is positive topographic openness; Upsl is upslope slope; FPL is flow path length; DI is downslope index; CA is catchment area; TWI is topographic wetness index; SPI is stream power index; LS is slope length factor. The values in bold have correlation coefficient > 0.5 and values in italic and bold indicate the highest correlation coefficient for soil properties.
*$P < 0.05$.
**$P < 0.005$.
***$P < 0.0001$.

Table 2.
Spearman's ranking correlation between topographic metrics and SOC density, ^{137}Cs inventory, and soil redistribution (SR) rates.

runoff shear stress increases, leading to enhance in sediment transport capacity, which increases the transports of ^{137}Cs and SOC-enriched fine fraction of sediments from low TR areas to the high TR areas.

Slope gradient (G) was another important factor for SOC density, ^{137}Cs inventory, and SR rates with absolute correlation coefficients larger than 0.6. Our findings are consistent with those of other researches, reporting high erosion rates in areas with relatively steep slopes [151, 155]. In agricultural fields, the main erosion processes include both water and tillage erosion [156–158]. Soil and associated SOC are transported to downslope due to gravity-driven lateral transport by overland and concentrated flows. Tillage operations would also cause redistribution of soil by small downslope movements of soil associated with each operation. Furthermore, as discussed in Section 2, G increase could enhance runoff and decrease infiltration, reducing water content in soil in the flat watershed area [15, 17, 18]. The controls of G on water and tillage erosion and water content could be related to the high slope impacts on soil properties in agricultural areas.

4.3.2 DTM-based models on soil property predictions

Since slope aspect (A) showed a weak correlation with SOC and no significant correlations with ^{137}Cs inventory and SR rate, we removed the A for the following DTM-based model development. Therefore, 17 topographic metrics, including slope gradient (G), curvature related metrics (P_Cur, Pl_Cur, and G_Cur), catchment area (CA), upslope slope (Upsl), downslope index (DI), flow path length (FPL), flow accumulation (FA), topographic relief principal components 1 and 2 (TRPC1 and TRPC2), topographic relief factors 1 and 2 (TRFA1 and TRFA2), positive topographic openness (PTO), topographic wetness index (TWI), stream power index (SPI), and slope length factor (LS) were used for building the MLR models. We only used TRPC1 and TRPC2 to represent topographic relief for MLR-PCA development and TRFA1 and TRFA2 for MLR-FAn.

The MLR, MLR-PCA, and MLR-FAn models were developed based on topographic and soil property data at the two field sites (**Table 3**). The MLR models showed the best simulations of SOC and SR rates with the highest R^2_{adj} and NSE values and the lowest RSR values over the three types of models. The two MLR models contained more than 7 predictors. The MLR-PCA model had a slightly lower efficiency than MLR-FAn model in simulating SOC density, but exhibited similar performance compared to the MLR-FAn model in SR rate simulations. The predictors included in the MLR-FAn were more than the MLR-PCA models. There were 6 and 5 factors included in MLR-FAn SOC and SR models, respectively; while only 4 and 5 components were contained in the MLR-PCA SOC and SR models, respectively.

Although the MLR showed the best simulation performance for the two field sites, the MLR-PCA had the highest prediction efficiency when applying models to predict the spatial patterns of SOC and SR rate over the watershed (**Figure 7a**). The SOC predictions by MLR-PCA explained 60% of the variability in observed SOC in the WCW. The NSE value was larger than 0.5 (0.591) and RSR value was <0.7 (0.639), which suggested a satisfactory performance of SOC prediction by the MLR-PCA model. The prediction efficiencies of MLR and MLR-FAn models were lower than the MLR-PCA model with correlation coefficients of 0.39 and 0.49, respectively. Based on these results, the SOC map over the watershed was generated based on the MLR-PCA model (**Figure 8**). The derived SOC map captured the majority spatial variability in SOC density as reflected by consistent spatial patterns between observed and simulated SOC density. High values of SOC density were observed in depressions and low values were found in ridges and sloping areas.

DOI: http://dx.doi.org/10.5772/intechopen.86109

	Model	R_{adj}^2	NSE	RSR
MLR				
SOC	2.98 + 0.071TRFA1 − 4.23G-9.29G_Cur + 0.0004FPL + 0.030TRPC2 + 0.103Pl_Cur + 0.063DI†	0.723	0.727	0.522
SR	2.12 − 3.12G + 0.107DI + 0.019TRFA2 + 0.0002FPL + 0.915Upsl + 0.010TRFA1 − 0.002SPI-1.53G_Cur	0.655	0.659	0.584
MLR-PCA				
SOC	2.94 − 0.060TPC2 − 0.024TPC3 - 0.051TPC7 + 0.037TPC1	0.684	0.686	0.560
SR	2.11 + 0.013TPC1 + 0.032TPC7 − 0.028TPC2−0.016TPC3 − 0.010TPC6	0.625	0.629	0.609
MLR-FAn				
SOC	2.92 − 0.101TFA1 + 0.074TFA4 + 0.045TFA7 + 0.026TFA3 + 0.037TFA2 − 0.027TFA3	0.706	0.710	0.538
SR	2.10−0.047TFA1 − 0.011TFA3 + 0.026TFA7 + 0.034TFA4 + 0.025TFA8	0.620	0.624	0.613

R^2_{adj} is adjusted coefficient of determination; NSE is Nash-Sutcliffeefficiency; RSR is ratioof the root mean square error (RMSE) to the standard deviation of measured data. TPC is topographic principal component; TFA is topographic factor; TRFA and TRPC are topographic relief factor and topographic relief component, respectively; G is slope gradient; Pl_Cur and G_Cur are plan curvature and general curvature, respectively; FPL is flow path length; DI is downslope index; Upsl is upslope slope; SPI is stream power index.
†The order of predictors is based on the stepwise selection procedure.

Table 3.
Models of soil organic carbon (SOC) density and soil redistribution (SR) rate based on cropland field sites 1 and 2.

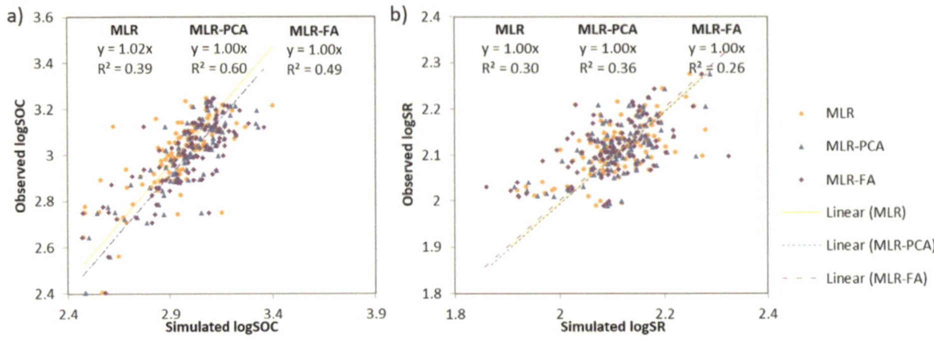

Figure 7.
Comparison of (a) soil organic carbon (SOC) density (kg m^{-2}) and (b) soil redistribution (SR) rate (t ha^{-1} $year^{-1}$) simulations to observations over the Walnut Creek watershed.

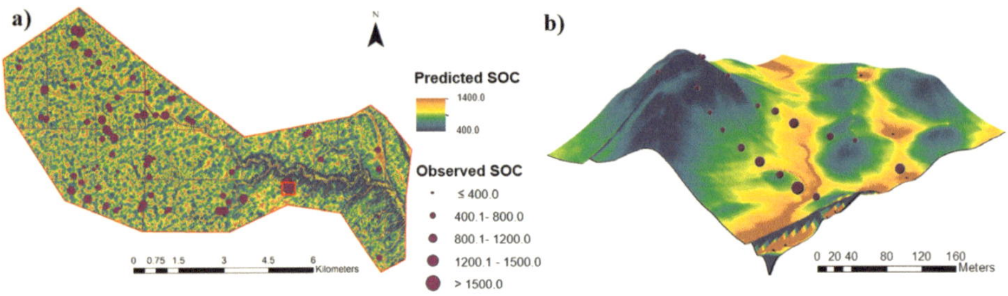

Figure 8.
Soil organic carbon (SOC) map obtained from the MLR-PCA model (a) within the Walnut Creek watershed and (b) along two transects.

Lower efficiencies were observed for SR rate than SOC density when compared the model predictions in the WCW (**Figure 7b**). The MLR-PCA SR model showed the highest correlation coefficient. However, the low NSE and high RSR values indicated that the model could not well predict SR rates when applying a model developed at field-scale for predictions at watershed-scale.

The better performance of MLR-PCA models relative to MRL models may be due to the exclusion of multicollinearity by PCA. High correlations ($|r| > 0.8$) were observed for some of topographic metrics, such as G and Upsl, and G_Cur and Pl_Cur. Uncertainty increases due to the high collinearity because models can be significantly influenced by small changes in the high collinearity predictors [159]. Thus, the MLR models were less stable with lower efficiencies in predicting SOC density and SR rates when applying to different spatial scales. The use of PCA could eliminate the multicollinearity and increase the stability of model since the PCA converted the 15-dimension topographic dataset to eight mutually independent combinations (TPCs) [159, 160].

Furthermore, by analysis of TPC loadings, hidden relationships between topographic metrics were uncovered, which could be another advantage of using the PCA [160]. For example, in this study, we selected TPCs 1, 2, 3, 6, and 7 for model development (**Table 4**). The high loading ($|\text{loading}| > 0.35$) topographic metric in TPC1 was G_Cur (−0.353), and thus, this component was associated with runoff divergence. G (0.475), TWI (−0.465), Upsl (0.419), and LS (0.396) were the high loading metrics for TPC2, which indicated that TPC2 was associated with soil water content. TPC3 were associated with runoff volume since the high loading metrics were FA (0.482), SPI (0.460), and CA (0.400). TPCs 6 and 7 were associated with

	TPC1(25%)	TPC2(24%)	TPC3(14%)	TPC6(5%)	TPC7(4%)
G	0.062	**0.475**	−0.035	−0.013	−0.183
P_Cur	−0.290	0.000	0.346	−0.070	−0.002
Pl_Cur	−0.283	0.107	−0.001	**0.485**	**0.383**
G_Cur	−**0.353**	0.054	0.275	0.025	0.100
FA	0.297	−0.042	**0.482**	0.179	0.131
TRPC1	0.309	−0.193	−0.237	0.113	−0.116
TRPC2	0.234	0.266	−0.118	0.084	**0.597**
PTO	−0.330	0.092	0.258	−0.292	0.217
Upsl	0.187	**0.419**	−0.143	−0.066	0.012
FPL	0.147	−0.168	−0.088	−**0.703**	**0.407**
DI	0.103	−0.220	−0.164	0.184	**0.435**
CA	0.326	−0.128	**0.400**	−0.160	−0.092
TWI	0.053	−**0.465**	−0.067	0.185	−0.047
SPI	0.345	−0.014	**0.460**	0.169	0.080
LS	0.256	**0.396**	0.050	0.011	−0.072

G is slope gradient; P_Cur, Pl_Cur, and G_Cur are profile curvature, plan curvature and general curvature, respectively; TRPC1 and TRPC2 are topographic relief components 1 and 2, respectively; PTO is positive topographic openness; Upsl is upslope slope; FPL is flow path length; DI is downslope index; CA is catchment area; TWI is topographic wetness index; SPI is stream power index; and LS is slope length factor.
The values in bold indicate loadings > 0.35.

Table 4.
Loadings in the selected topographic principal components (TPCs) calculating based on topographic metrics at the two field sites.

runoff velocity and flow acceleration, respectively. Based on the TPCs, we can obtain a better understanding of the controlling components for SOC distribution and SR. For the low-relief agriculture watershed under study, the spatial patterns of SOC and SR rate were mainly impacted by soil water content (TPC2) and runoff divergence (TPC1), respectively, according to the priority of TPCs used in model development. This conclusion is also consistent with findings by Fox and Papanicolaou [161] that indicated flow divergence significantly influenced soil erosion from uplands in a low-relief watershed.

The lower efficiencies of MLR-FAn than MLR-PCA may be because the latter approach diminishes the risk of over-fitting the models. The difference between PCA and FAn is that PCA considers all of the variance in the matrix, including unique, error and shared variance; while FAn extracts and exhibits shared variance only. Although some studies were preferable to FAn because of its ability to understand the underlying structure by extracting latent shared variance [162, 163], others also proved that there were almost no practical differences between the two methods [164, 165]. In this study, we found that both methods had similar performance during model calibration in small-scales. However, including more predictors in MLR-FAn models may enhance the instability of models and increase uncertainties during extrapolating prediction points over a large-scale [159, 166].

This case study demonstrated the importance of topography on soil properties in the low-relief watershed. DTM-based models are feasible for SOC predictions at different spatial scales. By combining MLR with PCA, the model efficiencies increased during soil property prediction. The DTM-based mapping techniques can be improved by further refinement remotely sensed data, improvement of the topographic dataset, and development of modeling techniques such as including Hybrid Regression and Artificial Intelligence techniques. The large-scale soil property maps can provide a more sound scientific basis for understanding of the mechanisms underlying the topographic impacts on soil movement in agricultural landscapes and the fate of SOC at the watershed and regional scales.

Acknowledgements

This research was supported by the USDA Natural Resources Conservation Service in association with the Wetland Component of the National Conservation Effects Assessment Project (NRCS 67-3A75-13-177). All data used in this study has been published on the USDA Ag Data Commons and can be accessed using DOI: 10.15482/USDA.ADC/1408098.

Conflict of interest

No conflict of interest exists in the submission manuscript.

Author details

Xia Li and Gregory W. McCarty*
Hydrology and Remote Sensing Laboratory, Beltsville Agricultural Research Center, U.S. Department of Agriculture-Agricultural Research Service, Beltsville, MD, USA

*Address all correspondence to: greg.mccarty@ars.usda.gov

References

[1] Moore ID, Grayson RB, Ladson DAR. Digital terrain modelling: A review of hydrological, geomorphological, and biological applications. Hydrological Processes. 1991;**5**:3-30

[2] Fox DM, Bryan RB. The relationship of soil loss by interrill erosion to slope gradient. Catena. 1999;**38**(3):211-222. DOI: 10.1016/S0341-8162(99)00072-7

[3] Pei T, Qin CZ, Zhu AX, Yang L, Luo M, Li B, et al. Mapping soil organic matter using the topographic wetness index: A comparative study based on different flow-direction algorithms and kriging methods. Ecological Indicators. 2010;**10**(3):610-619. DOI: 10.1016/j.ecolind.2009.10.005

[4] Li X, McCarty GW, Karlen DL, Cambardella CA. Topographic metric predictions of soil redistribution and organic carbon in Iowa cropland fields. Catena. 2018;**160**:222-232. DOI: 10.1016/j.catena.2017.09.026

[5] Speight JG. The role of topography in controlling throughflow generation. Earth Surface Processes. 1980;**5**:187-191. DOI: 10.1002/esp.3290030402

[6] Fissore C, Dalzell BJ, Berhe AA, Voegtle M, Evans M, Wu A. Influence of topography on soil organic carbon dynamics in a Southern California grassland. Catena. 2017;**149**:140-149. DOI: 10.1016/j.catena.2016.09.016

[7] Florinsky IV. Digital Terrain Analysis in Soil Science and Geology [Internet]. 2nd ed. Amsterdam: Academic Press; 2016. p. 486

[8] Walker PH, Hall GF, Protz R. Relation between landform parameters and soil properties. Soil Science Society of America Journal. 1968;**32**(1):101-104. DOI: 10.2136/sssaj1968.03615995003200010026x

[9] Aandahl AR. The characterization of slope positions and their influence on the total nitrogen content of a few virgin soils of western Iowa. Soil Science Society of America Journal. 1949;**13**:449-454. DOI: 10.2136/sssaj1949.036159950013000C0081x

[10] Skidmore AK. A comparison of techniques for calculating gradient and aspect from a gridded digital elevation model. International Journal of Geographical Information Systems. 1989;**3**(4):323-334. DOI: 10.1080/02693798908941519

[11] Doyle FJ. Digital terrain models: An overview. Photogrammetric Engineering and Remote Sensing. 1978;**44**(12):1481-1485

[12] Florinsky I, Eilers R, Manning G, Fuller L. Prediction of soil properties by digital terrain modelling. Environmental Modelling and Software. 2002;**17**(3):295-311. DOI: 10.1016/S1364-8152(01)00067-6

[13] Afshar FA, Ayoubi S, Jalalian A. Soil redistribution rate and its relationship with soil organic carbon and total nitrogen using ^{137}Cs technique in a cultivated complex hillslope in western Iran. Journal of Environmental Radioactivity. 2010;**101**(8):606-614. DOI: 10.1016/j.jenvrad.2010.03.008

[14] Florinsky IV, McMahon S, Burton DL. Topographic control of soil microbial activity: A case study of denitrifiers. Geoderma. 2004;**119**(1–2):33-53. DOI: 10.1016/S0016-7061(03)00224-6

[15] Huang J, Wu P, Zhao X. Effects of rainfall intensity, underlying surface and slope gradient on soil infiltration under simulated rainfall experiments. Catena. 2013;**104**:93-102. DOI: 10.1016/j.catena.2012.10.013

[16] Valentin C. Surface crusting in two alluvial soils of northen Niger. Geoderma. 1991;**48**:201-222

[17] Assouline S, Ben-Hur M. Effects of rainfall intensity and slope gradient on the dynamics of interrill erosion during soil surface sealing. Catena. 2006;**66**:211-220. DOI: 10.1016/j.catena.2006.02.005

[18] Janeau JL, Bricquet JP, Planchon O, Valentin C. Soil crusting and infiltration on steep slopes in northern Thailand. European Journal of Soil Science. 2003; **54**(3):543-553

[19] Radcliffe JE, Lefever KR. Aspect influences on pasture microclimate at Coopers Creek, North Canterbury. New Zealand Journal of Agricultural Research. 1981;**24**(1):55-66. DOI: 10.1080/00288233.1981.10420871

[20] Hanna AY, Harlan PW, Lewis DT. Soil available water as influenced by landscape position and aspect. Agronomy Journal. 1982;**74**(6): 999-1004. DOI: 10.2134/agronj1982.00021962007400060016x

[21] Rezaei SA, Gilkes RJ. The effects of landscape attributes and plant community on soil physical properties in rangelands. Geoderma. 2005;**125** (1–2):145-154. DOI: 10.1016/j.geoderma.2004.07.011

[22] Yimer F, Ledin S, Abdelkadir A. Soil organic carbon and total nitrogen stocks as affected by topographic aspect and vegetation in the Bale Mountains, Ethiopia. Geoderma. 2006;**135**:335-344. DOI: 10.1016/j.geoderma.2006.01.005

[23] Cerdá A. The influence of aspect and vegetation on seasonal changes in erosion under rainfall simulation on a clay soil in Spain. Canadian Journal of Soil Science. 1998;**78**(2):321-330

[24] Troch P, Van Loon E, Hilberts A. Analytical solutions to a hillslope-storage kinematic wave equation for subsurface flow. Advances in Water Resources. 2002;**25**(6):637-649. DOI: 10.1016/S0309-1708(02)00017-9

[25] Ritchie JC, McCarty GW, Venteris ER, Kaspar TC. Soil and soil organic carbon redistribution on the landscape. Geomorphology. 2007;**89**:163-171. DOI: 10.1016/j.geomorph.2006.07.021

[26] Li QY, Fang HY, Sun LY, Cai QG. Using the 137Cs technique to study the effect of soil redistribution on soil organic carbon and total nitrogen stocks in an agricultural catchment of Northeast China. Land Degradation & Development. 2014;**25**(4):350-359. DOI: 10.1002/ldr.2144

[27] Zaslavsky D, Rogowski AS. Hydrologic and morphologic implications of anistropy and infiltration in soil profile development. Soil Science Society of America Journal. 1969;**33**(4):594-599

[28] Pennock DJ, Zebarth BJ, de Jong E. Landform classification and soil distribution in hummocky terrain, Saskatchewan, Canada. Geoderma. 1987;**40**:297-315

[29] Kasai M, Marutani T, Reid LM, Trustrum NA. Estimation of temporally averaged sediment delivery ratio using aggradational terraces in headwater catchments of the Waipaoa River, North Island, New Zealand. Earth Surface Processes and Landforms. 2001;**26**(1): 1-16. DOI: 10.1002/1096-9837(200101)26:1<1::AID-ESP146>3.0.CO;2-9

[30] Wilson JP, Gallant JC. Digital terrain analysis. In: Wilson JP, Gallant JC, editors. Terrain Analysis: Principles and Applications. New York: John Wiley & Sons Ltd.; 2000. pp. 1-27

[31] Hjerdt KN. A new topographic index to quantify downslope controls on local drainage. Water Resources

Research. 2004;**40**(5):1-6. DOI: 10.1029/2004WR003130

[32] Seibert J, Stendahl J, Sørensen R. Topographical influences on soil properties in boreal forests. Geoderma. 2007;**141**(1–2):139-148. DOI: 10.1016/j.geoderma.2007.05.013

[33] Yanosek KA, Foltz RB, Dooley JH. Performance assessment of wood strand erosion control materials among varying slopes, soil textures, and cover amounts. Journal of Soil and Water Conservation. 2006;**61**(2):45-51

[34] Schubert J. Hydraulic aspects of riverbank filtration—Field studies. Journal of Hydrology. 2002;**266**:145-161

[35] Sharpley A, Kleinman P. Effect of rainfall simulator and plot scale on overland flow and phosphorus transport. Journal of Environmental Quality. 2003;**32**(6):2172-2179. DOI: 10.2134/jeq2003.2172

[36] Doody D, Moles R, Tunney H, Kurz I, Bourke D, Daly K, et al. Impact of flow path length and flow rate on phosphorus loss in simulated overland flow from a humic gleysol grassland soil. The Science of the Total Environment. 2006;**372**:247-255. DOI: 10.1016/j.scitotenv.2006.08.029

[37] Mitasova H, Hofierka J, Zlocha M, Iverson LR. Modeling topographic potential for erosion and deposition using GIS. International Journal of Geographical Information Systems. 1996;**10**(5):629-641. DOI: 10.1080/02693799608902101

[38] Zhang H, Yang Q, Li R, Liu Q, Moore D, He P, et al. Extension of a GIS procedure for calculating the RUSLE equation LS factor. Computers and Geosciences. 2013;**52**:177-188. DOI: 10.1016/j.cageo.2012.09.027

[39] Gessler PE, Chadwick OA, Chamran F, Althouse L, Holmes K. Modeling soil–landscape and ecosystem properties using terrain attributes. Soil Science Society of America Journal. 2000;**64**(6):2046-2056. DOI: 10.2136/sssaj2000.6462046x

[40] Dahal RK, Hasegawa S, Nonomura A, Yamanaka M, Masuda T, Nishino K. GIS-based weights-of-evidence modelling of rainfall-induced landslides in small catchments for landslide susceptibility mapping. Environmental Geology. 2008;**54**:311-324

[41] Summerfield MA, Hulton NJ. Natural controls of fluvial denudation rates in major world drainage basins. Journal of Geophysical Research. 1994;**99**(B7):13871-13883. DOI: 10.1029/94JB00715

[42] Tucker GE, Bras RL. Hillslope processes, drainage density, and landscape morphology. Water Resources Research. 1998;**34**(10):2751-2764

[43] Ahnert F. Functional relationships between denudation, relief, and uplift in large, mid-latitude drainage basins. American Journal of Science. 1970;**268**:243-263. DOI: 10.2475/ajs.268.3.243

[44] Yokoyama R, Shlrasawa M, Pike RJ. Visualizing topography by openness: A new application of image processing to digital elevation models. Photogrammetric Engineering and Remote Sensing. 2002;**68**(3):257-265

[45] Doneus M. Openness as visualization technique for interpretative mapping of airborne Lidar derived digital terrain models. Remote Sensing. 2013;**5**(12):6427-6442. DOI: 10.3390/rs5126427

[46] Li X, McCarty GW, Lang M, Ducey T, Hunt P, Miller J. Topographic and physicochemical controls on soil denitrification in prior converted croplands located on the Delmarva Peninsula, USA. Geoderma. 2018;**309**:

41-49. DOI: 10.1016/j.geoderma.2017.09.003

[47] Beven KJ, Kirkby MJ. A physically based, variable contributing area model of basin hydrology. Hydrological Sciences Bulletin. 1979;**24**(1):43-69. DOI: 10.1080/02626667909491834

[48] Lang MW, McCarty GW, Oesterling R, Yeo IY. Topographic metrics for improved mapping of forested wetlands. Wetlands. 2013; **33**(1):141-155. DOI: 10.1007/s13157-012-0359-8

[49] Lang MW, McCarty GW. Lidar intensity for improved detection of inundation below the forest canopy. Wetlands. 2009;**29**(4):1166-1178. DOI: 10.1672/08-197.1

[50] Conforti M, Aucelli PPC, Robustelli G, Scarciglia F. Geomorphology and GIS analysis for mapping gully erosion susceptibility in the Turbolo stream catchment (Northern Calabria, Italy). Natural Hazards. 2011;**56**(3):881-898. DOI: 10.1007/s11069-010-9598-2

[51] Tagil S, Jenness J. GIS-based automated landform classification and topographic, landcover and geologic attributes of landforms around the Yazoren Polje, Turkey. Journal of Applied Sciences. 2008;**8**(6):910-921

[52] Li X, McCarty GW. Use of principal components for scaling up topographic models to map soil redistribution and soil organic carbon. Journal of Visualized Experiments. 2018;**140**: e58189. DOI: 10.3791/58189

[53] Malone BP, McBratney AB, Minasny B, Laslett GM. Mapping continuous depth functions of soil carbon storage and available water capacity. Geoderma 2009;**154**(1–2):138-152. DOI: 10.1016/j.geoderma.2009.10.007

[54] Wischmeier WH, Smith DD. Predicting Rainfall Erosion Losses-a Guide to Conservation Planning. Washington, DC: U.S. Department of Agriculture; 1978. p. 537

[55] Florinsky IV. Combined analysis of digital terrain models and remotely sensed data in landscape investigations. Progress in Physical Geography. 1998;**22**(1):33-60. DOI: 10.1177/030913339802200102

[56] Angima SD, Stott DE, O'Neill MK, Ong CK, Weesies GA. Soil erosion prediction using RUSLE for central Kenyan highland conditions. Agriculture, Ecosystems and Environment. 2003;**97**(1–3):295-308. DOI: 10.1016/S0167-8809(03)00011-2

[57] Panagos P, Borrelli P, Meusburger K. A new European slope length and steepness factor (LS-factor) for modeling soil erosion by water. Geosciences. 2015;**5**(2):117-126. DOI: 10.3390/geosciences5020117

[58] Mehnatkesh A, Ayoubi S, Jalalian A, Sahrawat KL. Relationships between soil depth and terrain attributes in a semi arid hilly region in western Iran. Journal of Mountain Science. 2013;**10**(1): 163-172. DOI: 10.1007/s11629-013-2427-9

[59] Gessler PE, Moore ID, McKenzie NJ, Ryan PJ. Soil-landscape modelling and spatial prediction of soil attributes. International Journal of Geographical Information Systems. 1995;**9**(4):421-432. DOI: 10.1080/02693799508902047

[60] Park SJ, McSweeney KK, Lowery BB. Identification of the spatial distribution of soils using a process-based terrain characterization. Geoderma. 2001;**103**(3–4):249-272. DOI: 10.1016/S0016-7061(01)00042-8

[61] Chaplot V, Walter C. Subsurface topography to enhance the prediction of the spatial distribution of soil wetness.

Hydrological Processes. 2003;**17**(13): 2567-2580. DOI: 10.1002/hyp.1273

[62] Moore ID, Gessler PE, Nielsen GA, Peterson GA. Soil attribute prediction using terrain analysis. Soil Science Society of America Journal. 1993;**57**(2): 443-452. DOI: 10.2136/sssaj1993.572NPb

[63] Sumfleth K, Duttmann R. Prediction of soil property distribution in paddy soil landscapes using terrain data and satellite information as indicators. Ecological Indicators. 2008; **8**(5):485-501. DOI: 10.1016/j.ecolind.2007.05.005

[64] Vanwalleghem T, Poesen J, McBratney A, Deckers J. Spatial variability of soil horizon depth in natural loess-derived soils. Geoderma. 2010;**157**:37-45. DOI: 10.1016/j.geoderma.2010.03.013

[65] McBratney AB, Odeh IOA, Bishop TFA, Dunbar MS, Shatar TM. An overview of pedometric techniques for use in soil survey. Geoderma. 2000;**97** (3–4):293-327. DOI: 10.1016/S0016-7061(00)00043-4

[66] Bishop TFA, Mcbratney AB. A comparison of prediction methods for the creation of field-extent soil property maps. Geoderma. 2001;**103**:149-160

[67] Lark RM, Bishop TFA, Webster R. Using expert knowledge with control of false discovery rate to select regressors for prediction of soil properties. Geoderma. 2007;**138**(1–2):65-78. DOI: 10.1016/j.geoderma.2006.10.015

[68] Adhikari K, Hartemink AE, Minasny B, Bou Kheir R, Greve MB, Greve MH. Digital mapping of soil organic carbon contents and stocks in Denmark. PLoS One. 2014;**9**(8): e105519. DOI: 10.1371/journal.pone.0105519

[69] Li QQ, Yue TX, Wang CQ, Zhang WJ, Yu Y, Li B, et al. Spatially distributed modeling of soil organic matter across China: An application of artificial neural network approach. Catena. 2013;**104**:210-218. DOI: 10.1016/j.catena.2012.11.012

[70] Odeh IOA, McBratney AB, Chittleborough DJ. Spatial prediction of soil properties from landform attributes derived from a digital elevation model. Geoderma. 1994;**63**:197-214. DOI: 10.1016/0016-7061(94)90063-9

[71] Odeh IOA, McBratney AB, Chittleborough DJ. Further results on prediction of soil properties from terrain attributes: Heterotopic cokriging and regression-kriging. Geoderma. 1995;**67**: 215-226. DOI: 0016-7061/95/$09.50

[72] Zhu Q, Lin HS. Comparing ordinary kriging and regression kriging for soil properties in contrasting landscapes. Pedosphere. 2010;**20**(5):594-606. DOI: 10.1016/S1002-0160(10)60049-5

[73] Yao X, Fu B, Lü Y, Sun F, Wang S, Liu M. Comparison of four spatial interpolation methods for estimating soil moisture in a complex terrain catchment. PLoS One. 2013;**8**(1): e54660. DOI: 10.1371/journal.pone.0054660

[74] Kumar S, Lal R, Liu D. A geographically weighted regression kriging approach for mapping soil organic carbon stock. Geoderma. 2012; **189–190**:627-634. DOI: 10.1016/j.geoderma.2012.05.022

[75] Kravchenko AN, Robertson GP. Can topographical and yield data substantially improve total soil carbon mapping by regression kriging? Agronomy Journal. 2007;**99**(1):12-17. DOI: 10.2134/agronj2005.0251

[76] Bilgili AV. Spatial assessment of soil salinity in the Harran plain using multiple kriging techniques. Environmental Monitoring and

Assessment. 2013;**185**(1):777-795. DOI: 10.1007/s10661-012-2591-3

[77] Herbst M, Diekkrüger B, Vereecken H. Geostatistical co-regionalization of soil hydraulic properties in a micro-scale catchment using terrain attributes. Geoderma. 2006;**132**(1–2):206-221. DOI: 10.1016/j.geoderma.2005.05.008

[78] Motaghian HR, Mohammadi J. Spatial estimation of saturated hydraulic conductivity from terrain attributes using regression, kriging, and artificial neural networks. Pedosphere. 2011; **21**(2):170-177. DOI: 10.1016/ S1002-0160(11)60115-X

[79] Bourennane H, King D, Couturier A. Comparison of kriging with external drift and simple linear regression for predicting soil horizon thickness with different sample densities. Geoderma. 2000;**97**(3–4):255-271. DOI: 10.1016/ S0016-7061(00)00042-2

[80] Mirzaee S, Ghorbani-Dashtaki S, Mohammadi J, Asadi H, Asadzadeh F. Spatial variability of soil organic matter using remote sensing data. Catena. 2016;**145**:118-127. DOI: 10.1016/j. catena.2016.05.023

[81] Ersahin S. Comparing ordinary kriging and cokriging to estimate infiltration rate. Soil Science Society of America Journal. 2003;**67**:1848-1855. DOI: 10.2136/sssaj2003.1848

[82] Lagacherie P, Holmes S. Addressing geographical data errors in a classification tree for soil unit prediction. International Journal of Geographical Information Science. 1997; **11**(2):183-198. DOI: 10.1080/ 136588197242455

[83] Scull P, Franklin J, Chadwick OA. The application of classification tree analysis to soil type prediction in a desert landscape. Ecological Modelling. 2005;**181**(1):1-15. DOI: 10.1016/j. ecolmodel.2004.06.036

[84] Bui EN, Moran CJ. Disaggregation of polygons of surficial geology and soil maps using spatial modelling and legacy data. Geoderma. 2001;**103**(1–2):79-94. DOI: 10.1016/S0016-7061(01)00070-2

[85] Behrens T, Zhu AX, Schmidt K, Scholten T. Multi-scale digital terrain analysis and feature selection for digital soil mapping. Geoderma. 2010;**155** (3–4):175-185. DOI: 10.1016/j. geoderma.2009.07.010

[86] Grinand C, Arrouays D, Laroche B, Martin MP. Extrapolating regional soil landscapes from an existing soil map: Sampling intensity, validation procedures, and integration of spatial context. Geoderma. 2008;**143**(1–2): 180-190. DOI: 10.1016/j.geoderma. 2007.11.004

[87] Bin Z, Xin-gang Z, Ren-chao W. Automated soil resources mapping based on decision tree and Bayesian predictive modeling. Journal of Zhejiang University Science A. 2004;**5**(7): 782-795. DOI: 10.1631/jzus.2004.0782

[88] Geissen V, Kampichler C, López-de Llergo-Juárez JJ, Galindo-Acántara A. Superficial and subterranean soil erosion in Tabasco, tropical Mexico: Development of a decision tree modeling approach. Geoderma. 2007; **139**(3–4):277-287. DOI: 10.1016/j. geoderma.2007.01.002

[89] Odgers NP, Sun W, McBratney AB, Minasny B, Clifford D. Disaggregating and harmonising soil map units through resampled classification trees. Geoderma. 2014;**214–215**:91-100. DOI: 10.1016/j.geoderma.2013.09.024

[90] Nelson MA, Odeh IOA. Digital soil class mapping using legacy soil profile data: A comparison of a genetic algorithm and classification tree approach. Australian Journal of Soil Research. 2009;**47**(6):632-649. DOI: 10.1071/SR08224

[91] Sarmento EC, Giasson E, Weber EJ, Flores CA, Hasenack H. Disaggregating conventional soil maps with limited descriptive data: A knowledge-based approach in Serra Gaúcha, Brazil. Geoderma Regional. 2017;**8**:12-23. DOI: 10.1016/j.geodrs.2016.12.004

[92] Møller AB, Iversen BV, Beucher A, Greve MH. Prediction of soil drainage classes in Denmark by means of decision tree classification. Geoderma. 2017. (In Press). DOI: 10.1016/j.geoderma. 2017. 10.015

[93] Zhao Z, Chow TL, Yang Q, Rees HW, Benoy G, Xing Z, et al. Model prediction of soil drainage classes based on digital elevation model parameters and soil attributes from coarse resolution soil maps. Canadian Journal of Soil Science. 2008;**88**:787-799. DOI: 10.4141/CJSS08012

[94] Kidd DB, Malone BP, McBratney AB, Minasny B, Webb MA. Digital mapping of a soil drainage index for irrigated enterprise suitability in Tasmania, Australia. Soil Research. 2014;**52**(2):107-119. DOI: 10.1071/SR13100

[95] Lemercier B, Lacoste M, Loum M, Walter C. Extrapolation at regional scale of local soil knowledge using boosted classification trees: A two-step approach. Geoderma. 2012;**171–172**: 75-84. DOI: 10.1016/j.geoderma. 2011.03.010

[96] Beucher A, Møller AB, Greve MH. Artificial neural networks and decision tree classification for predicting soil drainage classes in Denmark. Geoderma. 2017. (In Press). DOI: 10.1016/j.geoderma. 2017.11.004

[97] Lark RM. Soil-landform relationships at within-field scales: An investigation using continuous classification. Geoderma. 1999;**92**(3–4): 141-165. DOI: 10.1016/S0016-7061(99) 00028-2

[98] Zhu AX, Hudson B, Burt J, Lubich K, Simonson D. Soil mapping using GIS, expert knowledge, and fuzzy logic. Soil Science Society of America Journal. 2001;**65**:1463-1472

[99] Zhu AX, Band LE, Dutton B, Nimlos TJ. Automated soil inference under fuzzy logic. Ecological Modelling. 1996; **90**(2):123-145. DOI: 10.1016/0304-3800 (95)00161-1

[100] Bui LV, Stahr K, Clemens G. A fuzzy logic slope-form system for predictive soil mapping of a landscape-scale area with strong relief conditions. Catena [Internet]. 2017;**155**:135-146. DOI: 10.1016/j.catena.2017.03.001

[101] Odeh IOA, Chittleborough DJ, McBratney AB. Soil pattern recognition with fuzzy-c-means: Application to classification and soil-landform interrelationships. Soil Science Society of America Journal. 1992;**56**(2):505-516. DOI: 10.2136/sssaj1992.0361599500 5600020027x

[102] Qi F, Zhu AX, Harrower M, Burt JE. Fuzzy soil mapping based on prototype category theory. Geoderma. 2006;**136**(3–4):774-787. DOI: 10.1016/j. geoderma.2006.06.001

[103] Akumu CE, Johnson JA, Etheridge D, Uhlig P, Woods M, Pitt DG, et al. GIS-fuzzy logic based approach in modeling soil texture: Using parts of the Clay Belt and Hornepayne region in Ontario Canada as a case study. Geoderma. 2015;**239–240**:13-24. DOI: 10.1016/j.geoderma. 2014.09.021

[104] Akumu CE, Woods M, Johnson JA, Pitt DG, Uhlig P, McMurray S. GIS-fuzzy logic technique in modeling soil depth classes: Using parts of the Clay Belt and Hornepayne region in Ontario, Canada as a case study. Geoderma. 2016;**283**:78-87. DOI: 10.1016/j. geoderma.2016.07.028

[105] de Bruin S, Stein A. Soil-landscape modelling using fuzzy c-means clustering of attribute data derived from a digital elevation model (DEM). Geoderma. 1998;**83**(1–2):17-33. DOI: 10.1016/S0016-7061(97)00143-2

[106] Meléndez-Pastor I, Pedreño JN, Lucas IG, Zorpas AA. A model for evaluating soil vulnerability to erosion using remote sensing data and a fuzzy logic system. In: Ramakrishnan S, editor. Modern Fuzzy Control Systems and its Applications. London: IntechOpen; 2017. pp. 437-453. DOI: 10.5772/32009

[107] Hengl T, Toomanian N, Reuter HI, Malakouti MJ. Methods to interpolate soil categorical variables from profile observations: Lessons from Iran. Geoderma. 2007;**140**(4):417-427. DOI: 10.1016/j.geoderma.2007.04.022

[108] Dobos E, Micheli E, Baumgardner MF, Biehl L, Helt T. Use of combined digital elevation model and satellite radiometric data for regional soil mapping. Geoderma. 2000;**97**:367-391

[109] Thomas AL, King D, Dambrine E, Couturier A, Roque J. Predicting soil classes with parameters derived from relief and geologic materials in a sandstone region of the Vosges mountains (Northeastern France). Geoderma. 1999;**90**(3–4):291-305. DOI: 10.1016/S0016-7061(98)00135-9

[110] Minasny B, McBratney AB. Incorporating taxonomic distance into spatial prediction and digital mapping of soil classes. Geoderma. 2007;**142**(3–4): 285-293. DOI: 10.1016/j.geoderma.2007.08.022

[111] Liu J, Pattey E, Nolin MC, Miller JR, Ka O. Mapping within-field soil drainage using remote sensing, DEM and apparent soil electrical conductivity. Geoderma. 2008;**143**(3–4):261-272. DOI: 10.1016/j.geoderma.2007.11.011

[112] Bell JC, Cunningham RL, Havens MW. Soil drainage class probability mapping using a soil-landscape model. Soil Science Society of America Journal. 1994;**58**(2):464-470. DOI: 10.2136/sssaj1994.03615995005800020031x

[113] Bell JC, Cunningham RL, Havens MW. Calibration and validation of a soil-landscape model for predicting soil drainage class. Soil Science Society of America Journal. 1992;**56**(6):1860-1866. DOI: 10.2136/sssaj1992.03615995005600060035x

[114] Kravchenko AN, Bollero GA, Omonode RA, Bullock DG. Quantitative mapping of soil drainage classes using topographical data and soil electrical conductivity. Soil Science Society of America Journal. 2002;**66**(1):235-243. DOI: 10.2136/sssaj2002.0235

[115] Nillsson NJ. Principles of Artifical Intelligence. San Francisco, CA, USA: Morgan Kaufmann; 2014. DOI: 10.1037/019640

[116] Russell SJ, Norvig P. Artificial Intelligence: A Modern Approach. 3rd ed. Upper Saddle River, NJ: Prentice Hall; 2010. DOI: 10.1016/B978-012161964-0/50009-1

[117] Lek S, Guégan JF. Artificial neural networks as a tool in ecological modelling, an introduction. Ecological Modelling. 1999;**120**:65-73. DOI: 10.1016/S0304-3800(99)00092-7

[118] Shichkin A, Buevich A, Sergeev A, Baglaeva E, Subbotina I, Vasilev J, et al. Training algorithms for artificial neural network in predicting of the content of chemical elements in the upper soil layer. AIP Conf. Proc. 2018;**2048**: 060004. DOI: 10.1063/1.5082119

[119] Basheer IA, Hajmeer M. Artificial neural networks: Fundamentals, computing, design, and application. Journal of Microbiological Methods.

2000;**43**:3-31. DOI: 10.12989/cac.2013.11.3.237

[120] Were K, Bui DT, Dick ØB, Singh BR. A comparative assessment of support vector regression, artificial neural networks, and random forests for predicting and mapping soil organic carbon stocks across an Afromontane landscape. Ecological Indicators. 2015; **52**:394-403. DOI: 10.1016/j.ecolind.2014.12.028

[121] Licznar P, Nearing MA. Artificial neural networks of soil erosion and runoff prediction at the plot scale. Catena. 2003;**51**:89-114

[122] Jafari A, Ayoubi S, Khademi H, Finke PA, Toomanian N. Selection of taxonomic level for soil mapping using integrated diversity indices and map purity data, a case study from an arid region of Iran. Geomorphology. 2013; **201**:86-97

[123] de Arruda GP, Demattê JAM, Chagas C da S, Fiorio PR, Souza AB, Fongaro CT. Digital soil mapping using reference area and artificial neural networks. Scientia Agricola. 2016;**73**(3): 266-273. DOI: 10.1590/0103-9016-2015-0131

[124] Freire S. Using artificial neural networks for digital soil mapping—A comparison of MLP and SOM approaches. In: AGILE 2013. Leuven. 2013

[125] de Carvalho Junior W, Chagas C da S, Fernandes Filho EI, Vieira CAO, Schaefer CEG, Bhering SB, et al. Digital soilscape mapping of tropical hillslope areas by neural networks. Scientia Agricola. 2011;**68**(6):691-696. DOI: 10.1590/S0103-90162011000600014

[126] Chagas C da S, Vieira CAO, Fernandes Filho EI. Comparison between artificial neural networks and maximum likelihood classification in digital soil mapping. Revista Brasileira de Ciência do Solo. 2013;**37**(2):339-351. DOI: 10.1590/S0100-06832013000200005

[127] Bagheri Bodaghabadi M, MartÍnez-Casasnovas J, Salehi MH, Mohammadi J, Esfandiarpoor Borujeni I, Toomanian N, et al. Digital soil mapping using artificial neural networks and terrain-related attributes. Pedosphere. 2015;**25**(4): 580-591. DOI: 10.1016/S1002-0160(15)30038-2

[128] Behrens T, Förster H, Scholten T, Steinrücken U, Spies ED, Goldschmitt M. Digital soil mapping using artificial neural networks. Journal of Plant Nutrition and Soil Science. 2005;**168**(1): 21-33. DOI: 10.1002/jpln.200421414

[129] Heung B, Ho HC, Zhang J, Knudby A, Bulmer CE, Schmidt MG. An overview and comparison of machine-learning techniques for classification purposes in digital soil mapping. Geoderma. 2016;**265**:62-77. DOI: 10.1016/j.geoderma.2015.11.014

[130] Schmidhuber J. Deep learning in neural networks: An overview. Neural Networks. 2015;**61**:85-117. DOI: 10.18388/abp.2015_1002

[131] Lecun Y, Bengio Y, Hinton G. Deep learning. Nature. 2015;**521**(7553): 436-444. DOI: 10.1038/nature14539

[132] Padarian J, Minasny B, McBratney AB. Using deep learning for digital soil mapping. Soil. 2019;**5**(1):79-89. DOI: 10.5194/soil-5-79-2019

[133] Lee H, Grosse R, Ranganath R, Ng AY. Convolutional deep belief networks for scalable unsupervised learning of hierarchical representation. In: Proceedings of the 26th Annual International Conference on Machine Learning. Montreal, Canada. 2009. pp. 609-616. DOI: 10.1145/2001269.2001295

[134] Song X, Zhang G, Liu F, Li D, Zhao Y, Yang J. Modeling spatio-temporal

distribution of soil moisture by deep learning-based cellular automata model. Journal of Arid Land. 2016;**8**(5): 734-748. DOI: 10.1007/s40333-016-0049-0

[135] Hafner D, Irpan A, Davidson J, Heess N. Learning hierarchical information flow with recurrent neural modules. In: Advances in Neural Information Processing Systems 30; 4-9 December 2017; Long Beach: NIPS; 2017. p. 6724-6733

[136] Carcano EC, Bartolini P, Muselli M, Piroddi L. Jordan recurrent neural network versus IHACRES in modelling daily stream flows. Journal of Hydrology. 2008;**362**(3–4):291-307. DOI: 10.1016/j.jhydrol.2008.08.026

[137] Díaz-Uriarte R, Alvarez de Andrés S. Gene selection and classification of microarray data using random forest. BMC Bioinformatics. 2006;7:1-13. DOI: 10.1186/1471-2105-7-3

[138] LEO B. Random Forest. Machine Learning. 2001;**45**(1):5-32. DOI: 10.1023/A:1010933404324

[139] Wiesmeier M, Barthold F, Blank B, Kögel-knabner I. Digital mapping of soil organic matter stocks using random Forest modeling in a semi-arid steppe ecosystem. Plant and Soil. 2011;**340**: 7-24. DOI: 10.1007/s11104-010-0425-z

[140] Grimm R, Behrens T, Märker M, Elsenbeer H. Soil organic carbon concentrations and stocks on Barro Colorado Island—Digital soil mapping using random forests analysis. Geoderma. 2008;**146**:102-113. DOI: 10.1016/j.geoderma.2008.05.008

[141] Hengl T, Heuvelink GBM, Kempen B, Leenaars JGB, Walsh MG, Shepherd KD, et al. Mapping soil properties of Africa at 250 m resolution: Random forests significantly improve current predictions. PLoS One. 2015;**10**(6): e0125814. DOI: 10.1371/journal.pone.0125814

[142] Zhang H, Wu P, Yin A, Yang X, Zhang M, Gao C. Prediction of soil organic carbon in an intensively managed reclamation zone of eastern China: A comparison of multiple linear regressions and the random forest model. The Science of the Total Environment. 2017;**592**:704-713

[143] Tesfa TK, Tarboton DG, Chandler DG, McNamara JP. Modeling soil depth from topographic and land cover attributes. Water Resources Research. 2009;**45**(10):1-16. DOI: 10.1029/2008WR007474

[144] Guo P, Li M, Luo W, Tang Q, Liu Z, Lin Z. Digital mapping of soil organic matter for rubber plantation at regional scale: An application of random forest plus residuals kriging approach. Geoderma. 2015;**237–238**: 49-59. DOI: 10.1016/j.geoderma.2014.08.009

[145] Rad MRP, Toomanian N, Khormali F, Brungard CW, Bayram C, Bogaert P. Updating soil survey maps using random forest and conditioned Latin hypercube sampling in the loess derived soils of northern Iran. Geoderma. 2014; **232–234**:97-106. DOI: 10.1016/j.geoderma.2014.04.036

[146] Heung B, Bulmer CE, Schmidt MG. Predictive soil parent material mapping at a regional-scale: A random Forest approach. Geoderma. 2014;**214–215**: 141-154. DOI: 10.1016/j.geoderma.2013.09.016

[147] Barthold FK, Wiesmeier M, Breuer L, Frede HG, Wu J, Blank FB. Land use and climate control the spatial distribution of soil types in the grasslands of Inner Mongolia. Journal of Arid Environments. 2013;**88**:194-205. DOI: 10.1016/j.jaridenv.2012.08.004

[148] Brungard CW, Boettinger JL, Duniway MC, Wills SA, Edwards TC. Machine learning for predicting soil classes in three semi-arid landscapes.

Geoderma. 2015;**239**:68-83. DOI: 10.1016/j.geoderma.2014.09.019

[149] Hatfield JL, Jaynes DB, Burkart MR, Cambardella CA, Moorman TB, Prueger JH, et al. Water quality in Walnut Creek watershed: Setting and farming practices. Journal of Environmental Quality. 1999;**28**(1): 11-24. DOI: 10.2134/jeq1999.00472425002800010002x

[150] Walling DE, Zhang Y, He Q. Models for deriving estimates of erosion and deposition rates from fallout radionuclide (caesium-137, excess lead-210, and beryllium-7) measurements and the development of user friendly software for model implementation. In: Impact of Soil Conservation Measures on Erosion Control and Soil Quality. IAEA-TECDOC-1665. 2011. pp. 11-33

[151] Young CJ, Liu S, Schumacher JA, Schumacher TE, Kaspar TC, McCarty GW, et al. Evaluation of a model framework to estimate soil and soil organic carbon redistribution by water and tillage using 137Cs in two U.S. Midwest agricultural fields. Geoderma. 2014;**232**:437-448. DOI: 10.1016/j.geoderma.2014.05.019

[152] Li X, Cui B, Yang Q, Lan Y. Impacts of water level fluctuations on detritus accumulation in Lake Baiyangdian, China. Ecohydrology. 2016;**9**(1):52-67. DOI: 10.1002/eco.1610

[153] Zhang G, Liu Y, Han Y, Zhang XC. Sediment transport and soil detachment on steep slopes: I. transport capacity estimation. Soil Science Society of America Journal. 2009;**73**(4):1291-1297. DOI: 10.2136/sssaj2008.0145

[154] Cogo NP, Moldenhauer WC, Foster GR. Effect of crop residue, tillage-induced roughness, and runoff velocity on size distribution of eroded soil aggregates. Soil Science Society of America Journal. 1983;**47**(5):1005-1008. DOI: 10.2136/sssaj1983.03615995004700050033x

[155] Du P, Walling DE. Using ^{137}Cs measurements to investigate the influence of erosion and soil redistribution on soil properties. Applied Radiation and Isotopes. 2011;**69**(5):717-726. DOI: 10.1016/j.apradiso.2011.01.022

[156] Venteris ER, McCarty GW, Ritchie JC, Gish T. Influence of management history and landscape variables on soil organic carbon and soil redistribution. Soil Science. 2004;**169**(11):787-795. DOI: 10.1097/01.ss.0000148742.75369.55

[157] Van Oost K, Govers G, de Alba S, Quine TA. Tillage erosion: A review of controlling factors and implications for soil quality. Progress in Physical Geography. 2006;**30**(4):443-466. DOI: 10.1191/0309133306pp487ra

[158] Heckrath G, Djurhuus J, Quine TA, Van Oost K, Govers G, Zhang Y. Tillage erosion and its effect on soil properties and crop yield in Denmark. Journal of Environmental Quality. 2005;**34**: 312-324. DOI: 10.1029/2002GB002010

[159] Dormann CF, Elith J, Bacher S, Buchmann C, Carl G, Carré G, et al. Collinearity: A review of methods to deal with it and a simulation study evaluating their performance. Ecography. 2013;**36**(1):027-046. DOI: 10.1111/j.1600-0587.2012.07348.x

[160] Roughgarden T, Valiant G. CS168: The modern algorithmic toolbox lecture #7: understanding principal component analysis (PCA). [Internet]. 2015. Available from: http://theory.stanford.edu/~tim/s15/l/l7.pdf

[161] Fox JF, Papanicolaou AN. The use of carbon and nitrogen isotopes to study watershed erosion processes. Journal of the American Water Resources Association. 2007;**43**(4):1047-1064. DOI: 10.1111/j.1752-1688.2007.00087.x

[162] Widaman K. Common factor analysis versus principal component

analysis: Differential bias in representing model parameters? Multivariate Behavioral Research. 1993; **28**(3):263-311. DOI: 10.1207/s15327906mbr2803

[163] Costello AB, Osborne JW. Best practices in exploratory factor analysis: Four recommendations for getting the most from your analysis. Practical Assessment, Research and Evaluation. 2005;**10**(7):1-9. DOI: 10.1109/IJCBS.2009.105

[164] De Winter JCF, Dodou D. Common factor analysis versus principal component analysis: A comparison of loadings by means of simulations. Communications in Statistics: Simulation and Computation. 2016;**45**(1):299-321. DOI: 10.1080/03610918.2013.862274

[165] Velicer W, Jackson D. Component analysis versus common factor analysis: Some further observations. Multivariate Behavioral Research. 1990;**25**(1):97-114. DOI: 10.1207/s15327906mbr2501

[166] Fodor IK. A survey of dimension reduction techniques. LLNL Technical Report, UCRL-ID-148494. [Internet]. 2002. Available from: http://citeseerx.ist.psu.edu/viewdoc/download?doi=10.1.1.122.201&rep=rep1&type=pdf

Chapter 2

Clay Minerals Mapping from Imaging Spectroscopy

Gilles Grandjean, Xavier Briottet, Karine Adeline, Anne Bourguignon and Audrey Hohmann

Abstract

Mapping subsurface clay minerals is an important issue because they have particular behaviors in terms of mechanics and hydrology that directly affects assets laid at the surface such as buildings, houses, etc. They have a direct impact in ground stability due to their swelling capacities, constraining infiltration processes during flooding, especially when moisture is important. So detecting and characterizing clay mineral in soils serve urban planning issues and improve the risk reduction by predicting impacts of subsidence on houses and infrastructures. High-resolution clay maps are thus needed with accurate indications on mineral species and abundances. Clay minerals, known as phyllosilicates, are divided in three main species: smectite, illite, and kaolinite. The smectite group highly contributes to the swelling behavior of soils, and because geotechnical soil analyses are expensive and time-consuming, it is urgent to develop new approaches for mapping clays' spatial distribution by using new technologies, e.g., ground spectrometer or remote hyperspectral cameras [0.4–2.5 μm]. These technics constitute efficient alternatives to conventional methods. We present in this chapter some recent results we got for characterizing clay species and their abundances from spectrometry, used either from a ground spectrometer or from hyperspectral cameras.

Keywords: clays, hyperspectral, spectroscopy, processing, unmixing

1. Introduction

Soils represent a complex environment, spatially and temporally dynamic in their structure as in their composition [1]. They provide essential services to humanity such as water storage and filtration, agriculture support, storage carbon to regulate the climate, and physical support of buildings. So, soil knowledge, in particular their clay mineral composition and their mapping, is necessary for the decision-making on the management of many human activities. The study of clay minerals is most of the time motivated by the assessment of the risk associated to shrinkage-swelling phenomenon that affects building; sometimes, they are also taken into consideration in flooding/infiltrating effects and in the evaluation of the vehicles' mobility. It is important to specify that the term "clay" may correspond to two distinct definitions in geology. From a physical point of view, clay minerals correspond to a texture class, e.g., a classification defined by the size of minerals in soils. In that classification, gravels are defined as elements larger than 2 mm, sands have a grain size of between 2 mm and 50 μm, silts have grain size between 50 and 2 μm, and clays have grain size lower than 2 μm.

From a mineralogical point of view, montmorillonite (i.e., the smectite group), illite, kaolinite, and interstratified minerals are the most common clay species that are commonly involved in swelling and shrinking processes. In the following, clays refer to this last mineralogical definition.

The shrinkage-swelling effect of soils is a phenomenon causing numerous damages on houses when built on soils containing smectite minerals. Indeed, these so-called swelling clays are sensitive to soil moisture content, since they shrink during periods of drought and swell after rain. The presence of water variations causes changes in volumes producing cracks in the soil structures and therefore vertical differential movements at the surface. In France, these damages reach 38% of natural disaster compensation costs after the floods. For the period 1990–2014, this overall cost represents a little more than 9 billion euros or 370 million euros per year [2]. In Great Britain, the association of insurers British estimated the cost of shrinkage-swelling to more than 400 million pounds each year [3]. In the USA, the economic cost of these claims is $15 billion annually [4]. As far as we know, population increase as well as projections of climate change should increase this risk at temperate latitudes, which in the future will affect areas previously untouched by drought. Identification of soils impacted by this phenomenon is currently based on specific mineral identifications, e.g., using X-ray diffraction (XRD) techniques, carried out on soil samples and difficult to implement at large scale. At the same time, some hazard maps (1:50000) were produced from geological data to identify clayed formations [5]. Unfortunately, these maps cannot consider local spatial heterogeneities, from one to hundreds of meters. In addition, mapping clay texture is not sufficient to evaluate the swelling capacity of clayed soils. To solve this issue, in situ and/or proximal sensors can be used.

Several authors have successfully quantified mineral clays in soils [6–8] by field spectroscopy and laboratory spectral measurements. In these studies, measurements were generally carried out on dry soils for avoiding spectral perturbations due to moisture, and under ideal conditions of illumination, away from real cases contexts found in the field. Airborne hyperspectral imagery has also been successfully used to detect clays [9–11], despite low spatial resolution offered by sensors and low signal-to-noise ratio in the spectral range affected by clays (1000–2500 nm). Recent advances in UAV-type platforms for hyperspectral imaging are expected to remove some of these limitations by a better spatial resolution of acquired images, moving from meters for airborne to centimeters for UAV [12, 13]. These advances must offer more pixels of pure soils and thus improve the quantification of clay minerals. Indeed, quantifying clay species from spectral data needs taking into account mixing spectral signatures of minerals, simply because they are mixed in the soil. Some studies hypothesize a linear mixture of soil mineral spectra (or "patchwork"). That means each component will have its spectral signature mixed in proportion to its abundance in the soil [6–8], which is an approximation because the diffusion of light induces nonlinearities on the spectral behavior of the reflectance present in an intimate mixture [14]. The impact of this phenomenon needs to be clearly assessed in order to correctly quantify clay species.

In the following, we propose a review on these different issues and describe the different approaches able to quantify clay species from hyperspectral data. This overview is based on different pieces of works realized in lab but also on the field, with different instrumental devices and several processing techniques.

2. Spectrometry experiments and data processing

The principle of spectrometry is based on the measurement of the interaction between an electromagnetic radiation and a given material at different frequencies. Applied to mineral characterization, this technique gives crystallo-chemical information on the material from its interaction with the incident radiation. Depending on the selected frequency of the radiation (ultraviolet, visible, infrared, etc.), the interaction produces various types of energy. This response is represented as a spectrum that is an intrinsic characteristic of the material [15]. The infrared radiation (IR) is an electromagnetic radiation, corresponding to the spectrum between 12,800 and 10 cm^{-1} (0.78–1000 μm). **Figure 1** shows the infrared electromagnetic spectrum that can be decomposed in three parts: the near, the middle, and the far IR. For mineral characterization, the domains of interest are the near-infrared (NIR) and the shortwave infrared (SWIR), which extend, respectively, from 0.75 to 1 μm and from 1 to 2.5 μm.

When an IR radiation interacts with a molecule, it can absorb partially and selectively this radiation, leading to modifications of the vibrational and rotational energy of the molecule. These energy losses lead to the presence of absorption bands at specific wavelengths corresponding to the frequencies at which the molecule is excited. The absorbed energy is therefore characteristic of each of the chemical bonds of the molecule. In the case of clay minerals, absorption bands are mostly visible in the SWIR domain. The complexity of working with absorption bands comes from the presence of water that also produces numerous absorption phenomena masking large parts of the spectrum (**Figure 2**). To predict soil properties related to the presence of clay minerals, intensive research has been carried out in reflectance spectroscopy in the visible near-infrared (VNIR; 300–1100 nm) and SWIR wavelength domains [16].

Interpreting correctly the spectrums resulting from interactions between SWIR irradiations and clayed soils is thus no straightforward, due to the noise coming from atmosphere, the presence of water molecules and the complexity of soils mineralogical composition.

Various approaches can be used to predict the clay mineralogical compositions of soils from measured spectra when the sample number is sufficiently high, e.g.,

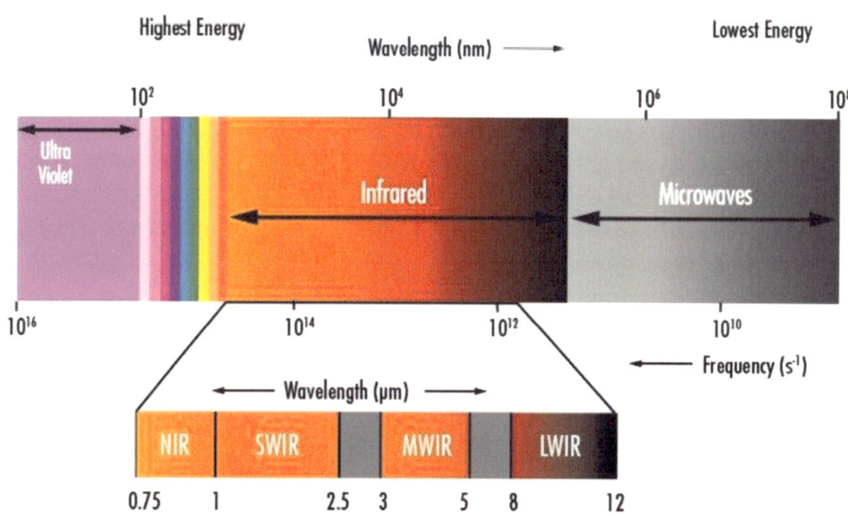

Figure 1.
Infrared domains expressed in terms of wavelengths.

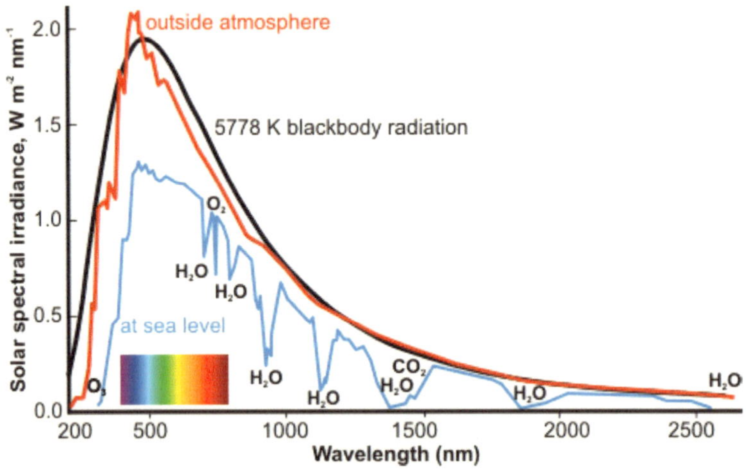

Figure 2.
Atmosphere and water absorption bands affecting the irradiated spectrum.

multivariable regression analysis (MRA) or partial least square regression (PLSR). For example, [17] have successfully estimated the smectite content of soils in the Colorado Front range by using a PLSR analysis of second derivative reflectance spectra measured in the field. MRA was also successfully used to quantify clay content in soil, independently of the nature of clay minerals [18, 19]. However, such approaches required a large number of observation samples to carry out the analysis but also to validate the regression accuracy. They are also site dependent, meaning that the calibration-validation processes need to be performed specifically for the studied sites. To tackle this issue with minimal uncertainties, we propose to start with simple experimental setups by analyzing in the laboratory the spectral responses of pure clays and mixtures of two or three species of clays.

2.1 Making and testing a spectral database from synthetic mixtures and a spectrometer

The objective of this first approach consists in preparing simple mixtures composed by pure clay minerals. They were prepared by [20] using the most common clays: montmorillonite, illite, and kaolinite, each of them provided by material sellers. The particle sizes of the minerals were measured with a VASCO-2 laser grain size analyzer and estimated to be about ~450 nm for the illite and the kaolinite and about ~475 nm for the smectite. The pure clay minerals were mixed using an agate mortar to produce mixed powders. A total of 27 binary mixtures of 10/90, 20/80, 30/70, 40/60, 50/50, 60/40, 70/30, 80/20, and 90/10 mass-percent ratios of kaolinite/illite, illite/montmorillonite, and montmorillonite/kaolinite were produced, as well as 19 ternary mixtures of kaolinite/illite/montmorillonite [20] (**Figure 3**).

All samples were dried and brought to humidity conditions of the laboratory. The reflectance spectra were measured in the laboratory using an ASD FieldSpec Pro. This spectrometer is portable and able to probe from 350 to 2500 nm in the electromagnetic spectrum. Its spectral resolution ranges from 10 nm with a 2 nm sampling interval in the SWIR. The mixtures were placed into Petri boxes, in contact with the probing system. A standard white Spectralon (Labsphere) was used to calibrate the reflectance reference. To increase the signal-to-noise ratio, the resulting spectrum was computed as the average of 10 spectral measurements [6].

As soon as the spectra are available for all the mixtures, a comparative analysis is used to relate a set of parametric observables derived from the spectrum morphology

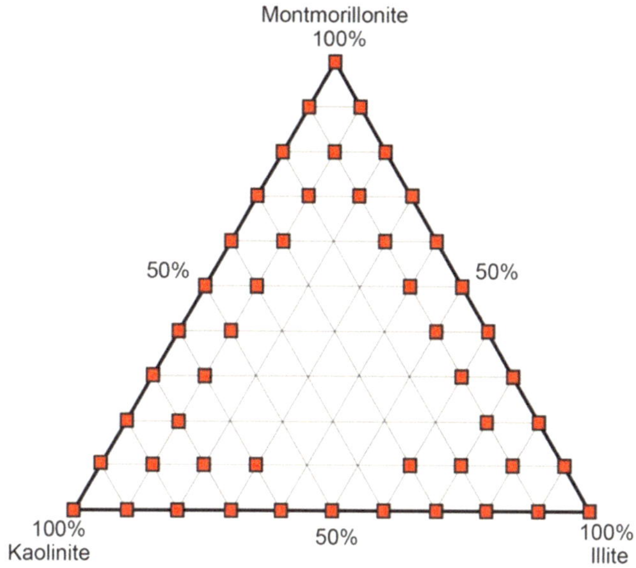

Figure 3.
Ternary diagram of kaolinite-illite-montmorillonite synthetic mixtures, modified from [20].

and the mineralogical composition of mixtures. Before this step, and in order to remove the large wavelength effects from each spectrum, a continuum-removal is applied as shown in **Figure 4** [14]. This processing leads to normalize the reflectance spectra and highlights absorption bands. The principle consists in connecting local maxima of the spectrum to obtain a good fit across the 350–2500 nm spectral domain [19]. After this processing, the continuum-removed spectrum has values ranging between 0 and 1 [18]. After this step, various geometrical parameters can be measured on the spectral curve as suggested by [21]. Indeed, this approach has the advantage to manipulate a few set of value to characterize a specific absorption band rather that considering overall values of the curve. The considered geometrical parameters are the following:

- The wavelength position corresponding to the minimum reflectance of the absorption band. In **Figure 4**, it corresponds to values around 1400 nm (P1400), 1900 nm (P1900), and 2200 nm (P2200).

- The depth, which is the length of the absorbing pattern along the reflectance axis. In **Figure 4**, the depth is estimated around 1400 nm (D1400), 1900 nm (D1900), and 2200 nm (D2200).

- The asymmetry of absorption band, calculated from the ratio between the right width and the left width measured at the half depth of the absorption band. In **Figure 4**, the asymmetry is about 1400 nm (A1400), 1900 nm (A1900), and 2200 nm (A2200).

- The width of the absorption band, measured at half depth. In **Figure 4**, the width is estimated to be around 1400 nm (W1400), 1900 nm (W1900), and 2200 nm (W2200).

As already mentioned by [21] or [22], the geometry of absorption bands around 1900 or 2200 nm is directly linked to the clay mineralogical composition. In particular, these studies show that the depth parameter can be efficiently used to assess the clay composition. If we plot the distribution of mixtures along 3 axes representing the depth parameter for 1400, 1900, and 2200 nm positions, we can identify regions where kaolinite, illite, and montmorillonite are particularly predominant, forming 3 corners of a triangular 3D shape. Elsewhere, kaolinite, illite, and montmorillonite contents in the mixtures decrease from their corner toward the opposite sides of the triangular shape [6] (**Figure 5**).

Even if these results are promising, they are not enough accurate to be exploited in real conditions. In particular, the development of a methodology able to statistically invert the abundance of clay species composing the mixtures from the absorption band parameters still needs to be tested. Such a study was carried out by [23], working with a higher complexity in preprocessing spectral data and trying to identify a robust unmixing method to estimate the clay abundances in the mixtures.

2.2 Processing laboratory hyperspectral images of synthetic mixtures, unmixing issues

To have a statistical assessment of the spectral response measured on the mixtures, the spectrometer was replaced by a hyperspectral optical sensor. This device is similar to that used by [24], with two cameras, located 1 m from the sample, and a lamp for each camera inclined to 35°. The reflected signal is recorded by two hyperspectral cameras (HySpex—Norsk Elektro Optikk—VNIR-1600 and SWIR-320 m-e). Only SWIR camera data is used, with 256 spectral bands and a spectral resolution of 6 nm in the range 1000–2500 nm. The camera has a measuring field of 240 mm (FOV 13.5°) and a spatial resolution of 0.75 mm. Between measurements, a white reference Spectralon R® is used to overcome any possible drift of instruments. Raw images highlight a nonuniformity of the illumination due to side effects. Experimental variograms realized on each band of reflectance images allowed to analyze this effect and to propose a masking protocol to remove

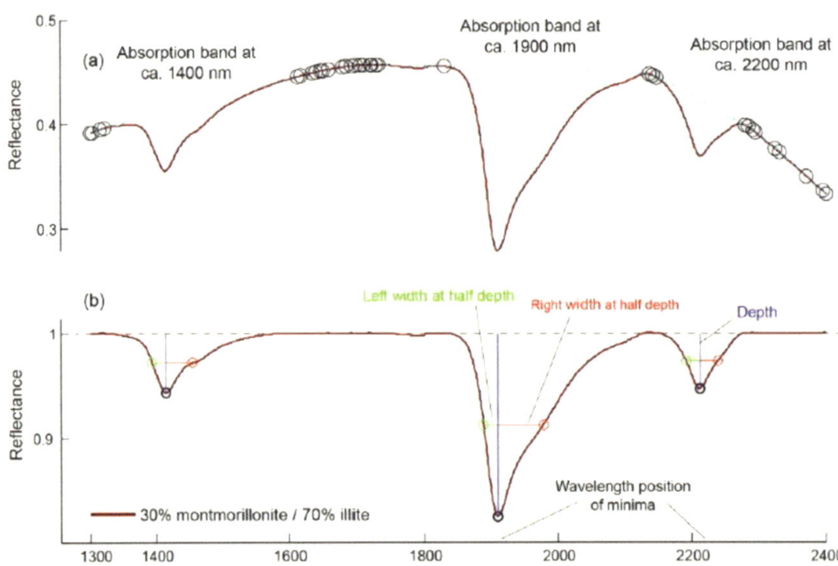

Figure 4.
Continuum-removal applied to a mixture spectrum of 30% montmorillonite and 70% illite. (a) spectrum before Continuum-removal, (b) Geometrical parameters used to characterize the absorption bands: location of the minimum (black circles), depth (blue line), left width at half depth (green line), and right width at half depth (red line).

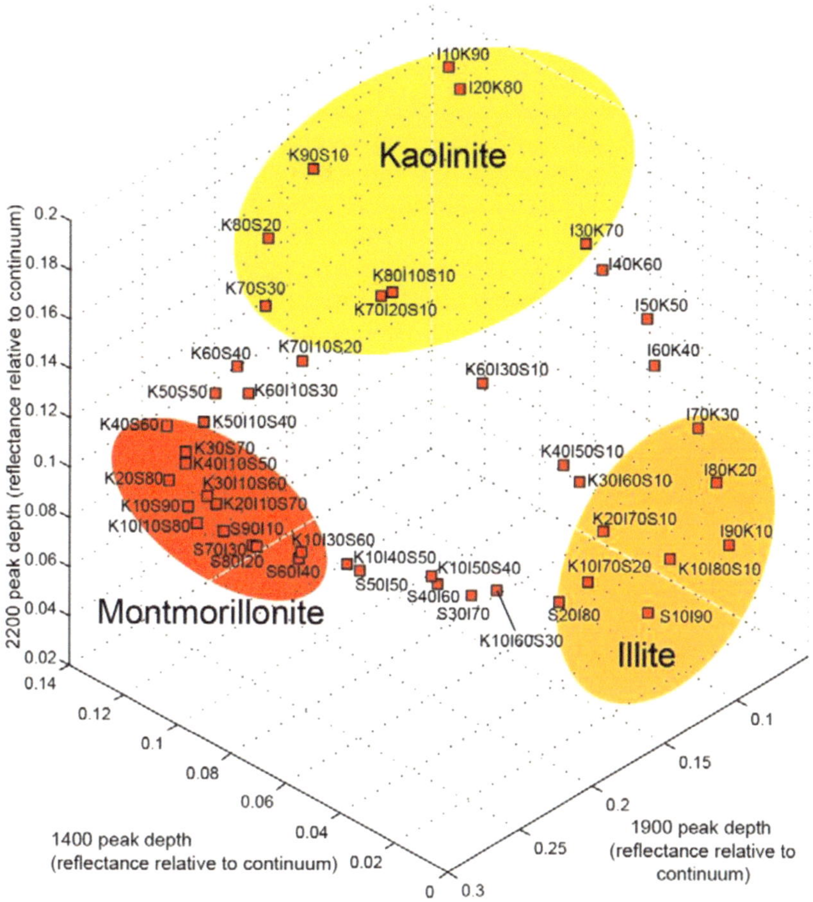

Figure 5.
3D-diagram showing the distribution of the synthetic mixtures according to the depth parameter. From [6].

pixels too far from the homogeneous behavior observed at the center of images. The following methodological chain is based on (i) spectral preprocessing to transform reflectance spectra in a standardized form and (ii) linear and nonlinear unmixing algorithms to derive mineral abundance for each mixture (**Figure 6**). Preprocessing techniques were selected from the literature and concern:

- Standard normal variate (SNV) consists in applying a translation and a homothety of the spectrum using its mean and standard deviation [25].

- Continuum-removal (CR) deletes the continuum to normalize the reflectance spectrum [26].

- Continuous wavelet transform (CWT) splits the signal into a wavelet sum of Gaussian function (e.g., "Mexican Hat"). The signal is broken down into 10 scales, the first one (corresponding to the noise) and scales higher than 5 (global variations of the spectrum-continuum) are suppressed [27].

- Hapke's model [28] estimates the single diffusion albedo considering that the medium is an isotropic mixture with the same particle size for all components.

- First derivative (1St SGD) calculated according to [29].

- Transformation into pseudo-absorbance (Log (1/R)) based on the correlation between the bands of spectral absorption and concentration of compounds [25].

Once spectra are preprocessed, several unmixing techniques can be tested to determine abundances. Before, it is necessary to compare observed spectrum to reference spectrum, i.e., spectrum of pure minerals (end-members) present in the mixture. On the one hand, if all the minerals present are known, one can use spectral libraries existing in the literature. Otherwise, algorithms able to determine in the observed data those which represent the most pure end-members can be used such as SISAL [30] or minimum volume [31]. Four linear and nonlinear unmixing algorithms were used to estimate abundances in clay minerals from mixtures described in the previous chapter (**Figure** 7):

- FCLS is the most popular linear unmixing method and has nonnegativity constraints (abundances must be equal or higher than 0), and the sum of abundances of each end-member must equal to one [32].
- MESMA, similar to FCLS, takes into account the intra-class variability of each mixing pole.
- The GBM method [33] can take into account nonlinear effects by the way of an additional parameter.
- The multilinear model (MLM) method [34] uses a parameter to manage nonlinearity; for zero, the model becomes linear.

The results show that the unmixing method performance depends on the mineralogy of the mixture, the difficulty arising when clay species have very similar spectrum in the considered wavelengths. We can also note that the linear and nonlinear methods have similar performances on these mixtures, the recommended method being in fact the simplest to use, i.e., FCLS. Finally, the benefit brought by

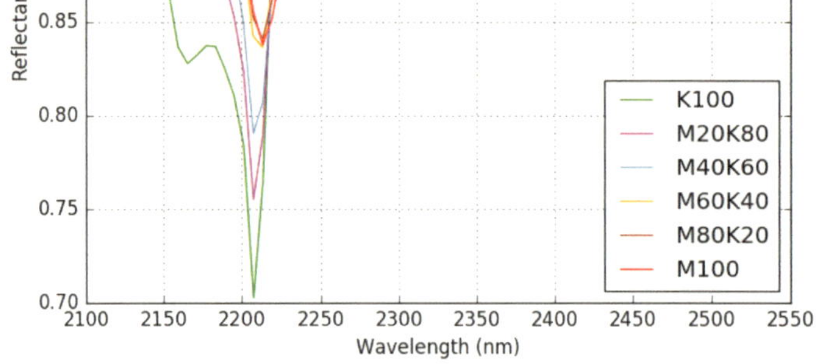

Figure 6.
Mean spectra of hyperspectral images after continuum-removal correction for different montmorillonite/kaolinite abundances.

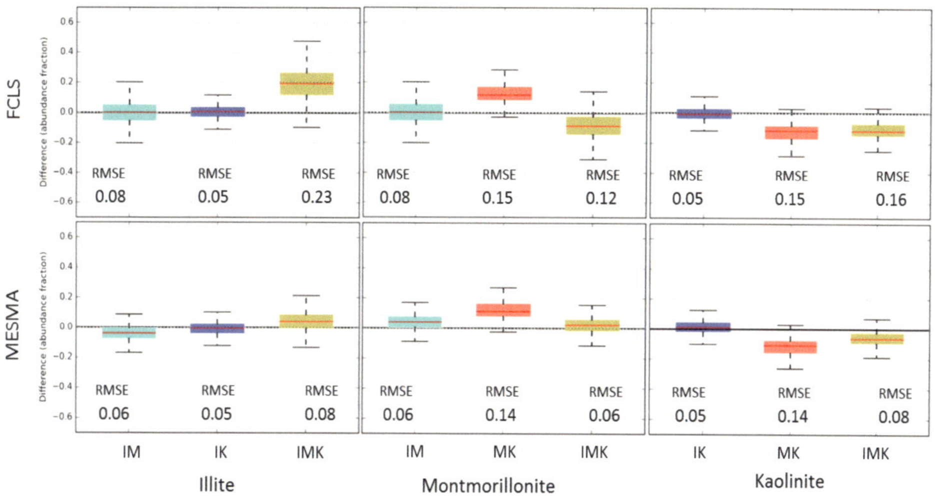

Figure 7.
Variability of abundances predictions decreases with MESMA and FCLS, where I stands for illite, K for kaolinite, and M for montmorillonite.

spectral preprocessing is very important. CWT and first SGD give one of the best performances on unmixing quality by decreasing the intra-sample variability [35].

2.3 From lab measurements to field observations

A good example of validation and comparison between lab models and field observations is given by [36]. The sampling area is located close to Orleans city (France) along the Loire River. The fluvial deposits are mainly composed of sandy materials contained in a clay matrix, containing also pebbles and boulders. In this study, 332 samples of soil were collected, spread over the various geological formations where swelling risk is present. As in [6], spectrum where decomposed in geometrical parameters, more suitable for quantitative analyses. As shown in **Figure 8a**, the ratios of the depth parameters for different absorption bands (D1400 over D2200 vs. D1900 over D2200) demonstrate that the montmorillonite and illite end-members appear in the scattered plot. This approach could be used to roughly evaluate the content of these clay species in the soil samples.

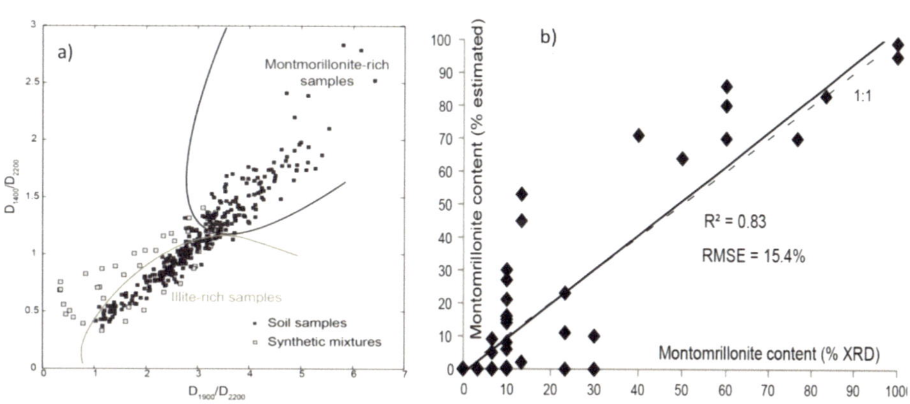

Figure 8.
(a) Scattered plot of studied samples represented according to two ratios of depth parameters; (b) correlation between montmorillonite content measured from XRD and estimated from spectroscopy [36].

To evaluate the uncertainties related to this approach, 31 samples of the dataset were analyzed using X-ray diffraction, and comparison were carried out between montmorillonite content measured from XRD and montmorillonite content estimated from spectroscopy. Although the distribution of points presents a certain dispersion, the correlation ratio, close to 0.84, confirms the potential of using geometrical characteristics of spectra to assess the abundance of clay species.

3. Conclusion

The geotechnical issues raised by swelling clays need to be addressed to evaluate the vulnerability of buildings and houses lying on clayed soils geological environment. To reduce costs of analyses, classically consisting in lab measurements (e.g., XRD), methodologies based on spectroscopy can be used. This chapter shows last advances in evaluating clay species abundances, in particular for montmorillonite, from spectroscopy or hyperspectral approaches in the SWIR domain.

A first step was the development of metrics to discriminate clay minerals from their spectral response. For this purpose, mixtures were realized from pure clay minerals, and their spectra were systematically analyzed using geometrical parameter such as the depth of the different absorbing band patterns. From this database, we showed that a discrimination was possible, at least to have a qualitative estimation of the swelling capacities of concerned soils. This result was validated from the field by comparing the abundances estimated coming from spectroscopy and from XRD techniques. Another approach based on hyperspectral image processing was presented. Different preprocessing algorithms and unmixing techniques were applied to the mixture dataset for performance evaluation. The results are also very conclusive since RMS values between estimated and observed abundances are satisfactory.

This overview gives important perspective in the domain. If spectroscopy can evaluate clay mineral abundances in soils and in particular those who have swelling capacities, the possibility to use remote hyperspectral camera for this purpose could be considered. The next perspective are thus to test this probing technique to field data in real condition. The heterogeneous solar lightning; the presence of vegetation, calcite, or quartz pebbles; and possibility of moisture variations in soils are, for instance, the next issues to work on. Due to recent developments in UAV, new possibilities could be found for carrying hyperspectral cameras in SWIR domain and reaching information with higher signal-to-noise ratio and better resolution. These advances should open new perspectives for accurate and less expensive productions of clay maps.

Acknowledgment

Authors would like to thank BRGM and ONERA for funding these studies. We also thank the different students (C. Truche, G. Duffrechou, and E. Ducasse) who take in charge a large part of experiments, analyses, and processing, as well as the technicians involved in the lab tasks.

Author details

Gilles Grandjean[1*], Xavier Briottet[2], Karine Adeline[2], Anne Bourguignon[1] and Audrey Hohmann[1]

1 BRGM, Orléans, France

2 ONERA, Toulouse Center, Toulouse University, France

*Address all correspondence to: g.grandjean@brgm.fr

References

[1] Carmon N, Ben-Dor E. An advanced analytical approach for spectral-based modelling of soil properties. International Journal of Emerging Technology and Advanced Engineering. 2017;**7**(3):90-97

[2] Corti T, Muccione V, Köllner-Heck P, Bresch D, Seneviratne SI. Simulating past droughts and associated building damages in France. Hydrology and Earth System Sciences. 2009;**13**:1739-1747

[3] Crilly MS, Driscoll RMC. The behaviour of lightly loaded piles in swelling ground and implications for their design. Proceedings of the Institution of Civil Engineers: Geotechnical Engineering. 2000;**143**(1):3-16

[4] Nelson J, Miller DJ. Expansive Soils: Problems and Practice in Foundation and Pavement Engineering. Professional Series. NewYork: Wiley; 1992. ISBN 9780471181149

[5] Bouchut J, Giot D. Cartographie de l'aléa retrait-gonflement des argiles dans le département du Loiret. Rapport technique RP-53316-FR, BRGM. 2004. Available from: http://infoterre.brgm.fr/rapports/RP-53316-FR.pdf

[6] Dufréchou V, Grandjean G, Bourguignon A. Geometrical analysis of laboratory soil spectra in the short-wave infrared domain: Clay composition and estimation of the swelling potential. Geoderma. 2015;**243-244**(2015):92-107

[7] Mulder VL, Plötze M, de Bruin S, Schaepman ME, Mavris C, Kokaly RF, et al. Quantifying mineral abundances of complex mixtures by coupling spectral deconvolution of SWIR spectra (2.1-2.4 mm) and regression tree analysis. Geoderma. 2013;**207-208**:279-290

[8] Viscarra Rossel RA, McGlynn RN, McBratney AB. Determining the composition of mineral-organic mixes using UV-vis-NIR diffuse reflectance spectroscopy. Geoderma. 2006;**137**(1-2):70-82

[9] Bedini E, van der Meer F, van Ruitenbeek F. Use of HyMap imaging spectrometer data to map mineralogy in the Rodalquilar caldera, Southeast Spain. International Journal of Remote Sensing. 2009;**30**(2):327-348

[10] Chabrillat S, Goetz AFH, Krosley L, Wolsen H. Use of hyperspectral images in the identification and mapping of expansive clay soils and the role of spatial resolution. Remote Sensing of Environment. 2002;**82**(2-3):431-445

[11] Kruze FA, Boardman JW, Huntington JF. Comparison of airborne hyperspectral data and eo-1 hyperion for mineral mapping. IEEE Transactions on Geoscience and Remote Sensing. 2003;**41**(6):1388-1400

[12] van der Meer FD, van der Werff HMA, van Ruitenbeek FJA, Hecker CA, Bakker WH, Noomen MF, et al. Multi- and hyperspectral geologic remote sensing: A review. International Journal of Applied Earth Observation and Geoinformation. 2012;**14**(1):112-128

[13] Jackisch R, Lorenz S, Zimmermann R, Möckel R, Gloaguen R. Drone-borne hyperspectral monitoring of acid mine drainage: An example from the sokolov lignite district. Remote Sensing. 2018;**10**(3):385

[14] Clark RN. Spectroscopy of rocks and minerals, and principles of spectroscopy. In: Rencz AN, editor. Remote Sensing for the Earth Sciences—Manual of Remote Sensing. Vol. 3. New York: JohnWiley and Sons in

cooperation with the American Society for Photogrammetry and Remote Sensing; 1999. pp. 3-58

[15] King PL, Swayze GA. Infrared Spectroscopy in Geochemistry, Exploration Geochemistry, and Remote Sensing. Québec: Mineralogical Association of Canada; 2004

[16] Bourguignon A, Delpont G, Chevrel S, Chabrillat S. Detection and mapping of shrink–swell clays in SW France, using ASTER imagery. Geological Society, London, Special Publications. 2007;**283**:117-124

[17] Goetz AFH, Chabrillat S, Lu Z. Field reflectance spectrometry for detection of swelling clays at construction sites. Field Analytical Chemistry and Technology. 2001;**5**(3):143-155

[18] Gomez C, Lagacherie P, Coulouma G. Continuum removal versus PLSR method for clay and calcium carbonate content estimation from laboratory and airborne hyperspectral measurements. Geoderma. 2008;**148**:141-148

[19] Viscarra Rossel RA, Cattle SR, Ortega A, Fouad Y. In situ measurements of soil colour, mineral composition and clay content by Vis–NIR spectroscopy. Geoderma. 2009;**150**:253-266

[20] Truche C. Caractérisation et quantification des minéraux argileux dans les sols expansifs par spectroscopie infrarouge aux échelles du laboratoire et du terrain. Toulouse, France: Laboratoire des Mécanismes et Transfert en Géologie. Université Paul Sabatier — Toulouse III; 2011. p. 229

[21] van der Meer F. Analysis of spectral absorption features in hyperspectral imagery. International Journal of Applied Earth Observation and Geoinformation. 2004;**5**:55-68

[22] Kariuki PC, Van DerMeer F, Siderius W. Classification of soils based on engineering indices and spectral data. International Journal of Remote Sensing. 2003;**24**:2567-2574

[23] Ducasse E, Hohmann A, Adeline K, Oltra-Carrio R, Bourguignon A, et al. Clay minerals abundances estimation under hyperspectral laboratory data. In: 10th EARSeL Workshop on Imaging Spectroscopy. Zurich, Switzerland; 2017

[24] Adar S, Shkolnisky Y, Ben-Dor E. Change detection of soils under small-scale laboratory conditions using imaging spectroscopy sensors. Geoderma. 2014;**216**:19-29

[25] Esquerre C, Gowen AA, Burger J, Downey G, O'Donnell CP. Suppressing sample morphology effects in near infrared spectral imaging using chemometric data pre-treatments. Chemometrics and Intelligent Laboratory Systems. 2012;**117**:129-137

[26] Clark RN, Roush TL. Reflectance spectroscopy: Quantitative analysis techniques for remote sensing applications. Journal of Geophysical Research—Solid Earth. 1984;**89**(B7):6329-6340

[27] Feng J, Rogge D, Rivard B. Comparison of lithological mapping results from airborne hyperspectral VNIR-SWIR, LWIR and combined data. International Journal of Applied Earth Observation and Geoinformation. 2017;**64**:340-353

[28] Hapke B. Bidirectional reflectance spectroscopy: 1. Theory. Journal of Geophysical Research—Solid Earth. 1981;**86**(B4):3039-3054

[29] Savitzky A, Golay MJE. Smoothing and differentiation of data by simplified least squares procedures. Analytical Chemistry. 1964;**36**:1627-1639, 1964

[30] Bioucas-Dias JM, Plaza A, Dobigeon N, Parente M, Du Q, Gader P, et al. Hyperspectral unmixing overview: Geometrical, statistical, and sparse

regression-based approaches. IEEE Journal of Selected Topics in Applied Earth Observations and Remote Sensing. 2012;**5**(2):354-379

[31] Craig MD. Minimum-volume transforms for remotely sensed data. IEEE Transactions on Geoscience and Remote Sensing. 1994;**32**(3):542-552

[32] Heinz DC, Chang CI. Fully constrained least squares linear spectral mixture analysis method for material quantification in hyperspectral imagery. IEEE Transactions on Geoscience and Remote Sensing. 2001;**39**(3):529-545

[33] Halimi A, Altmann Y, Dobigeon N, Tourneret JY. Nonlinear unmixing of hyperspectral images using a generalized bilinear model. IEEE Transactions on Geoscience and Remote Sensing. 2011;**49**(11):4153-4162

[34] Heylen R, Scheunders P. A multilinear mixing model for nonlinear spectral unmixing. IEEE Transactions on Geoscience and Remote Sensing. 2016;**54**(1):240-251

[35] Ducasse E. Cartographie fine de l'argile minéralogique par démélange d'images hyperspectrales à trés haute résolution spatiale [thesis]. Univ. Toulouse. 2019

[36] Dufréchou G, Hohmann A, Bourguignon A, Grandjean G. Targeting and mapping expansive soils (Loiret, France): Geometrical analysis of laboratory soil spectra in the short-wave infrared domain (1100-2500 nm). Bulletin de la Société Géologique de France. 2016;**187**(3):169-181

Chapter 3

The Impact of Land Use and Land Cover Changes on the Nkula Dam in the Middle Shire River Catchment, Malawi

Maureen Kapute Mzuza, Weiguo Zhang, Fanuel Kapute and Xiaodao Wei

Abstract

Land use and land cover changes over a 26-year period for the middle Shire River catchment, Malawi, in southern Africa, were assessed using geographic information systems (GIS) and remote sensing techniques. The catchment area under study was divided into two sections, western and eastern sides of the Shire River. High rate of deforestation averaging 4.3% per annum was observed and more pronounced in the western side of the river. Rapid population growth and increase in gross domestic product (GDP) are identified as the major drivers of deforestation and forest degradation due to clearing of vast fields for agriculture, land expansion for urban settlement, and cutting down of trees for wood fuel energy. Deforestation in the middle Shire River catchment has resulted into increased soil loss through erosion causing huge accumulation of sediment at the Nkula B Hydroelectric Power Dam downstream and, consequently, causing serious problems with generation of hydroelectricity. Frequent droughts and floods in the area have drastically affected crop production forcing people into cutting down of trees for charcoal as a livelihood strategy. Combined techniques such as GIS, remote sensing, and socioeconomic factors used in this study could be applied in other places where similar challenges occur.

Keywords: LUCC, GIS, remote sensing, soil, Malawi

1. Introduction

Land use and land cover changes have significant environmental consequences at local, regional, and global scales. These changes have intense implications at the regional and global scales for global loss of biodiversity, distresses in hydrological cycles, increase in soil erosion, and sediment loads [1]. At the local level, changes in the use of land and its cover affect watershed runoff, microclimatic resources, processes of land degradation and landscape-level biodiversity, soil erosion, and sediment loads [2]. All these have direct impacts on livelihoods of local societies.

The Shire River in Malawi, southern Africa, is among the areas where land use land cover change (LUCC) has become more prevalent in recent years resulting into severe soil erosion and causing heavy siltation downstream [3–9]. The river is an important source of livelihood to many people, using the water for agriculture, domestic purposes, and the generation of electricity [6, 8, 10]. One of the most important structures across the Shire River is the Nkula B Hydroelectric Power Station situated in the middle section of the river. The dam at Nkula Falls that supplies water into the power station has, in recent times, been threatened with massive siltation, some studies attributing this to increased human population and agricultural activities [5, 6, 8]. The conceptual setting of this study originates from a strong link that exists between land use change and soil erosion [8, 11–15]. Land use and management practices are important factors in determining the extent of soil erosion [8, 15]. Good vegetation cover promotes infiltration of water into the ground and soil retention, while deforestation results into increased runoff than infiltration occurring during periods of more precipitation [16–18]. Increased runoff consequently leads to stronger soil erosion usually in areas with poor vegetation cover [8, 19–20]. Erosion of soil under continuous cultivation is the most serious form of resource degradation occurring in Malawi [3, 8, 19, 21–23]. The rate of soil loss in Malawi is currently estimated at 29 t/ha/year [24], which is higher than the previously reported 20 t/ha/year [21]. In the middle Shire River, estimated soil loss between the year 2000 and 2014 ranged from 0.1 to 21.1 t/ha/year [24, 25]. According to the Malawi Government Report (2015), the middle Shire River catchment has many bright spots (areas experiencing high soil loss but declining trends over time), for example, Neno and Ntcheu in the west and Zomba and Chiradzulu in the eastern side of the river.

The question regarding land use changes over time, and its driving forces in the middle Shire River catchment nevertheless remain unresolved [4, 6]. Such knowledge is critical to the development of policies and action plans necessary for changing current LUCC trends in the area as it has been observed in other places [26–30]. Furthermore, problems of LUCC are global and serious in many developing countries where increasing population has resulted into excessive pressure on natural resources [8, 30].

The study was carried out to understand the impact of land use and land cover changes on the Nkula Dam in the middle Shire River catchment, Malawi. The LUCC drivers analyzed in this study include biophysical changes (e.g., climate change) and human activities (e.g., population, poverty, land policies, and GDP growth) [3, 4, 6]. Climate and socioeconomic data were compiled to analyze the drivers of LUCC in the study area. Geographic information systems (GIS) and remote sensing techniques which are gaining increased recognition globally as rapid methods of acquiring and analyzing up-to-date information over a large geographical area were used in the study [30–33].

2. Study area and methods

2.1 Description of the study area

The Shire River is the largest river in Malawi, originating from Lake Malawi which supports vast agricultural and socioeconomic activities in its catchment (**Figure 1**) [34]. The river is divided into three sections, namely, the upper, middle, and lower Shire [34, 35]. This study focused on the catchment of the middle section of the river which includes the Shire Plain which is bounded by mountains on both sides and the Nkula Dam downstream [34, 36]. The plain is more extensive to the west of the river than it is to the east (**Figure 1**). The middle section of the Shire River has eight administrative districts, supporting a population of about 5 million people (**Figure 1**).

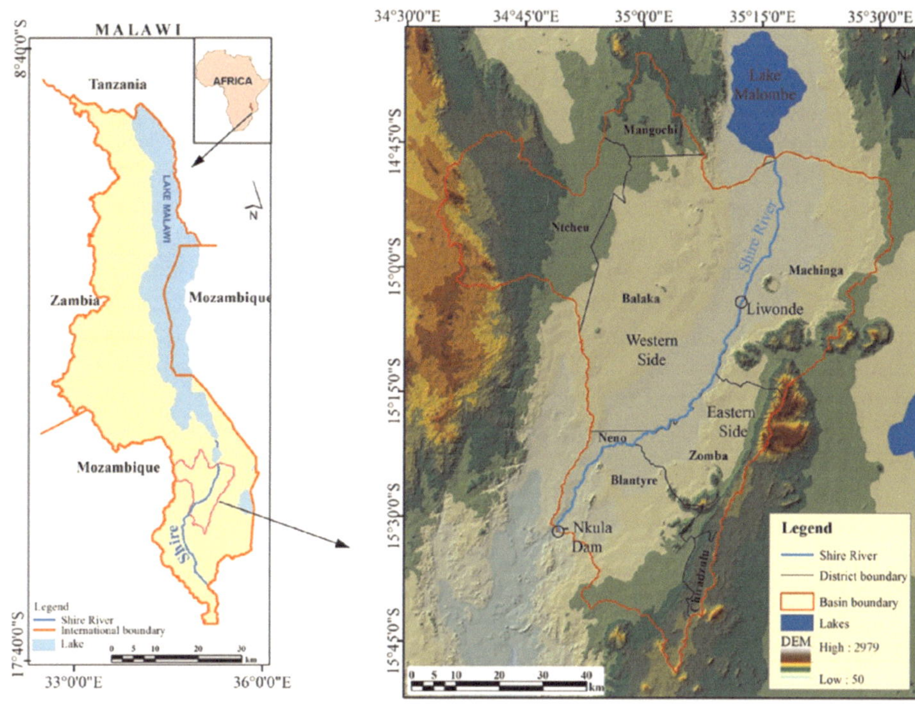

Figure 1.
Map of Malawi (left) showing the middle Shire River and its catchment (right). Eight administrative districts are located in the study area.

Climate in the middle Shire River catchment area varies due to differences in altitude with annual average precipitation ranging from 750 to 2500 mm [35, 37]. Highlands receive more rain which begins in November and ends late in April [6, 37]. Annual average temperature of the area is around 23°C, with highlands in the east experiencing cooler temperatures than plains in the west [6, 35]. The rocks in the study area are mainly composed of Precambrian basement complex and igneous rocks [37]. Amphibolite and granulite facies are dominant in the western and eastern side of the Shire River, respectively, while soils in the river's catchment are dominated by Cambisols [6, 24, 37].

2.2 Data collection procedure

The following procedures were followed in order to answer the study questions: firstly, six Landsat images for the dry seasons (to avoid cloud cover effects) of 1989, 1993, 2000, 2006, 2011, and 2015 were downloaded from the United States Geological Survey (USGS, http://glovis.usgs.gov/) at Level 1 T using different paths and rows (167/070, 167/071, 168/070, and 168/071). All images had a spatial resolution of 30 m which is large enough to visualize changes in land use [38] from Landsat 5, 7, and 8. Secondly, meteorological, topographical, and socioeconomic data from 1989 to 2015 were collected from the Malawi Department of Meteorological Services and Statistics [24, 36]. The third stage was the processing of the Landsat images and, finally, classification of land use which was followed by analysis of different land covers. Statistical analysis was done on data for the topography of the catchment area, temperature, rainfall, population, and GDP in order to determine drivers of LUCC.

2.2.1 Remote sensing image processing

Landsat images were processed using ENVI 5.1 Software to study information on the types of land use and their spatial patterns. To analyze these spatial patterns, the following steps were followed: firstly, relative radiometric correction was done on each band to eliminate errors arising from radiation caused by weather conditions; secondly, multiband combination of Landsat images was done in preparation for research spectral characteristics of various types of land use; thirdly, geometric correction of remote sensing images was done using Malawi DEM, Universal Transverse Mercator Projection, Arc 1960, and UTM Zone 36S, based on 1:50,000 topographic map scale so that it fits with the Landsat images [38, 39]. This helps to eliminate position errors of Landsat images which the terrain, position of the sun, and angle sensor may produce. A mosaic of required images was prepared and a single image generated. Atmospheric Landsat images were then corrected by ENVI 5.1 FLAASH module.

2.2.2 Land use classification

After processing the Landsat images, identification of different land use classes was done where some visual designs like texture, tone, and the effect zones were used [38]. The land in the study area was classified according to its use or description such as cultivated land, water, forest (indigenous and plantations were combined), etc. When identifying the training sites, the spectral signatures separability of all the eight land use classes presented in **Table 1** were verified including control fields in situ that were also set for validation of each classified image [38]. Land use types were classified by supervised classification maximum likelihood method since it's among the broadly used methods in the scientific literature in addition to it being the fastest and easy to use and giving a perfect interpretation of the outcomes [38–44]. In addition, the method is able to accommodate covarying data which is common with satellite image data [41, 45]. Representative zones for each desired class were located in the image with adequate number of pixels covering the known classes to reduce the image noise [38]. Secondly, training area number and percentage were identified in order to classify several training and test areas. These results were compared with supporting ground data so that the new training statistics could be derived. Thirdly, a statistical file known as spectral signature was created by the image processing software for each class because each and every pixel can only be assigned to one spectral class. Lastly, each pixel was allocated to the most likely class based on the maximum likelihood algorithm where each pixel is assigned to the spectral class that has the greatest probability density function for the multispectral values of the pixel. Maximum likelihood algorithm is the most commonly used algorithm in which a pixel is classified into the corresponding class [38, 43, 46]. Land cover types were then classified into the following eight main classes according to Anderson et al. [47]: (1) forest, (2) shrubland, (3) grassland, (4) cultivated land, (5) bare land, (6) water bodies, (7) wetland, and (8) artificial surfaces (**Table 1**).

A total of 165 training sites (sampled portions of the scene, purposely selected, for the derivation of the training statistics) were chosen for each image to ensure that all spectral classes constituting each land use and land cover categories were adequately represented in the training statistics to classify the entire scene [48]. Classification was done using ground checkpoints, digital topographic maps, vegetation cover map, and the researchers' knowledge of the study area [49, 50]. A total of 156 sampling points (GPS + photograph) were collected out of the 165 training sites during the dry season to avoid cloud cover effects which is more common in

No.	Land cover class	Description
1	Forest	Woodland open general (15–65%) with herbaceous layer. Broadleaved deciduous trees, closed >(70–60)%. Vegetative cover is in balance with the abiotic and biotic forces of its biotope
2	Shrubland	Closed to open (thicket) (15–100%) scattered trees
3	Grassland	Herbaceous closed vegetation (15–100%) with some trees, shrub Savannah, and permanent marsh
4	Cultivated land	Areas where the natural vegetation has been removed or modified and replaced by other types of vegetative cover of anthropogenic origin. All vegetation that is planted or cultivated with intent to harvest is included in this class
5	Bare land	Bare rock and/or coarse fragments. Areas that do not have an artificial cover as a result of human activities. These areas include areas with less than 4% vegetative cover
6	Water bodies	This class refers to areas that are naturally covered by water, such as lakes, rivers, snow, or ice
7	Wetlands	Areas that are transitional between pure terrestrial and aquatic systems and where the water table is usually at or near the surface or the land is covered by shallow water
8	Artificial surfaces	Areas that have an artificial cover as a result of human activities, such as construction (cities, towns, and transportation), extraction (open mines and quarries), or waste disposal

Table 1.
Land cover classes considered and their description [71].

rainy season. Land use types at the sampling sites were evaluated according to field surveys (photographs + GPS) where photographs were taken using a camera and coordinates of the spot were taken using GPS. Accuracy of the supervised classification methods was checked by a confusion matrix of accuracy (**Table 2**) [38, 44, 51] to ensure that various measures, such as error-rate, accuracy, specificity, sensitivity, and precision, were checked.

Landsat image classified type results were compared with the field survey results to evaluate their accuracy and then calculated using confusion matrix evaluation table (**Table 2**).

2.2.3 Statistical analysis

LUCC drivers were mainly analyzed using descriptive methods due to inavailability of spatial socioeconomic data from the government database. Pearson correlation coefficients between socioeconomic data and land use types were analyzed in SPSS for Windows version 10.

3. Results

3.1 Land use and land cover changes over the past 26 years

The overall classification accuracy ranged from 82 to 94% (**Table 2**). The western side of the Shire River covers an area of approximately 3353 km^2, while the eastern side is 2770 km^2 comprising 55 and 45% of the total area, respectively. Regions were defined by slope of less than 10o as plain/flat area. According to **Table 3**, total plain/flat area covers 2417 km^2 which is lesser compared to highlands

Actual type	Classified type									
	Forest	Shrubland	Grassland	Cultivated land	Artificial surfaces	Wetland	Water bodies	Bare land	Actual sum	Accuracy
Forest	9	1	1	0	0	0	0	0	11	82%
Shrubland	0	14	1	1	0	0	0	0	16	88%
Grassland	0	1	20	1	0	1	0	1	24	83%
Cultivated land	0	0	1	21	1	0	0	0	23	91%
Artificial surfaces	1	1	1	2	34	0	0	0	39	87%
Wetland	0	0	1	0	0	8	0	0	9	89%
Water bodies	0	1	0	0	0	1	32	0	34	94%
Bare land	0	0	0	0	0	0	0	0	0	0
Classified sum	10	18	25	25	35	10	32	1	156	

Table 2.
Confusion matrix of accuracy evaluation in middle Shire River catchment in 2015.

Area/coverage	Plain (≤10°)		Highlands (10–90°)	
	Area (km^2)	Percentage (%)	Area (km^2)	Percentage (%)
Western side	1429	59	1075	29
Eastern side	988	41	2631	71
Total catchment area	2417	100	3706	100

Table 3.
Distribution of plains and highlands in eastern and western side of the middle Shire River.

(with slope ranging from 10o to 90o) covering 3706 km^2. Eastern and western plain/flat areas cover 988 and 1429 km^2, representing 41 and 59% of the total plain/flat area of the study area, respectively (**Table 3**).

The middle Shire River catchment is dominated by shrubland, grassland, cultivated land, and forestland, which accounted for 36, 28, 22, and 12% in 1989, respectively (**Figure 2**).

Findings (**Table 4**) show significant land use and land cover changes in the middle Shire River catchment over the 26-year period.

Artificial and cultivated land increased by 65 and 52%, respectively, in the 26-year period, while forest cover, grass, and shrubland decreased by 35, 27, and 7%, respectively. Other land classes such as wetlands and water bodies show

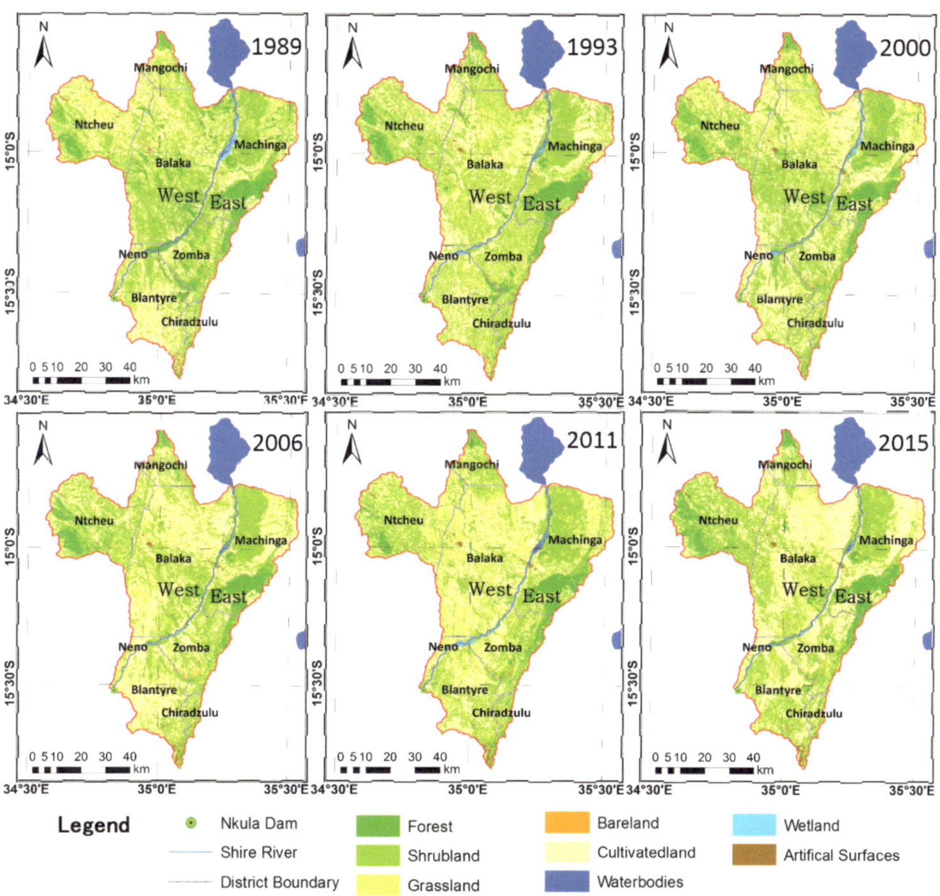

Figure 2.
Land use and land cover changes from 1989 to 2015.

Land cover type	Year											
	1989		1993		2000		2006		2011		2015	
	Area (km^2)	%	Area (km^2)	%	Area (km^2)	%	Area (km^2)	%	Area (km^2)	%	Area (km^2)	%
Forest	739	12.07	679	11.08	545	8.90	479	7.82	481	7.86	662	10.80
Shrubland	2201	35.95	1986	32.44	2264	36.97	2043	33.37	1835	31.85	2040	32.97
Grassland	1719	28.07	1838	30.02	1451	23.69	1692	27.63	1617	24.53	1255	20.52
Cultivated land	1367	22.33	1538	25.12	1745	28.50	1814	29.64	2067	33.76	2073	34.09
Artificial surfaces	26	0.43	28	0.45	33	0.54	37	0.60	39	0.64	43	0.71
Wetland	35	0.57	23	0.38	56	0.91	19	0.31	38	0.63	20	0.34
Water bodies	31	0.51	30	0.49	20	0.33	30	0.49	36	0.58	22	0.44
Bare land	4	0.06	2	0.03	9	0.15	9	0.15	9	0.15	8	0.13

Table 4.
Area (km^2) and percentages of different land cover types from the year 1989 to 2015.

fluctuations (**Figure 2** and **Table 4**). Spatially, in 1989, total cultivated land in the western side was 694 km^2 which increased to 1226 km^2 by the year 2015, representing 21 and 37% of the total land in the western side, respectively (**Table 5**).

This suggests an increase of 16% of cultivated land in the western side between 1989 and 2015. In the eastern side, cultivated land increased from 673 to 862 km^2 within the same period, representing 24 and 31%, respectively, of the total land area indicating a 7% change. In 1989, the western side of the Shire River catchment mainly consisted of shrubland, grassland, and forestland which accounted for 35, 33, and 10%, respectively. In the eastern side, shrubland, grassland, and forestland accounted for 37, 22, and 15%, respectively. The western side (Balaka, Neno, and Ntcheu) and eastern side (Zomba) are the main districts where forest, shrubland, and grassland decreased the most. For example, in Balaka District, forest area reduced from 11% in 1989 to 2% in 2011 before increasing to 3% in 2015, while shrubland decreased from 38% in 1989 to 18% in 2011 and then increased to 23% in 2015. Forestland in Neno District decreased from 10% in 1989 to 1% in 2011 and then increased up to 5% in 2015, while shrubland decreased from 35% in 1989 to 19% in 2015 and grassland from 27% in 1989 to 17% in 2015 with some fluctuations in between the years. In Ntcheu District, grassland decreased from 35% in 1989 to 15% in 2015. Forest cover in Zomba district declined from 19% in 1989 to 7% in 2006 and then started to increase from 2011 reaching 12% in 2015. Shrubland decreased from 41% in 1989 to 27% in 2015 in the same district.

3.2 Changes in climate, population, and GDP

Results indicate some fluctuations in the amount of rainfall received in the area within the 26-year period that might be due to climate change as a result of land use and land cover changes due to human activities (**Figure 3**).

Rainfall in the catchment area declined continuously from 1989 to 1993, culminating into the drought of 1992 and 1993 (**Figure 3**) [52, 53]. Malawi is regularly affected by drought and floods [53]. The country (including the study area) was affected by heavy floods in 1989, 1998, 2000, 2001, and 2015, destroying crops and displacing many people (**Figure 3**) [53]. Earlier studies indicate that rainy season in Malawi is dominated by tropical and extratropical influences with links to the El

Location/district		Year					
		1989	1993	2000	2006	2011	2015
Western side	Balaka	335	556	627	655	688	853
	Mangochi	59	51	41	80	47	91
	Neno	25	41	49	38	28	53
	Ntcheu	275	298	219	226	219	228
	Total area	694	946	935	999	982	1226
Eastern side	Blantyre	359	264	362	381	244	278
	Chiradzulu	33	9	19	17	18	23
	Machinga	184	247	264	263	135	368
	Zomba	96	71	165	155	122	194
	Total area	673	591	810	816	520	862

Table 5.
Changes in cultivated land area (km^2) in districts of the middle Shire River catchment.

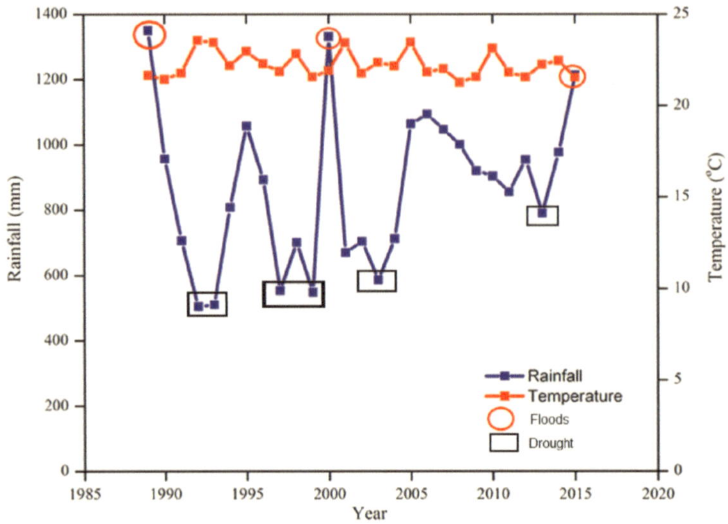

Figure 3.
Annual rainfall and temperature for the middle Shire River catchment from 1989 to 2015. Circles represent flood years, while rectangles represent drought years (Source: Malawi Meteorological Department).

Niño-Southern Oscillation (ENSO) [54, 55]. Actually, this is reported for the whole of Southern Africa [56].

The population of Malawi which includes districts under study on the western (Mangochi, Balaka, Ntcheu, and Neno) and eastern sides of the middle Shire River (Blantyre, Zomba, Machinga, and Chiradzulu) has been increasing steadily since the 1980s (**Figure 4**).

Increased population is more pronounced in urban areas. For example, in 2015, Blantyre and Zomba cities had 3006 and 2240 people per km^2, respectively [34, 53, 57]. There has been a general increase in the GDP over the past 26 years especially between 2006 and 2011 and falling between 1993 and 2003 (**Figure 4**).

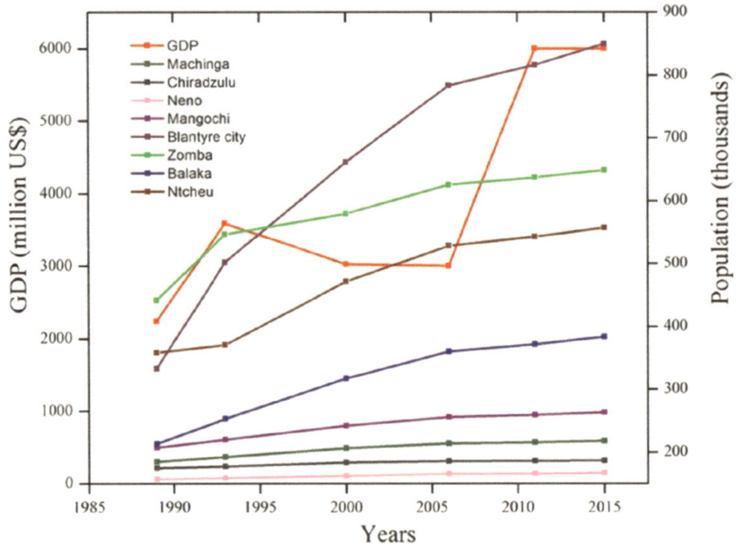

Figure 4.
Population of districts in the middle Shire River catchment area and GDP (US$) for Malawi from 1989 to 2015 [53].

4. Discussion

4.1 Drivers of LUCC in the middle Shire River catchment

Rainfall affects LUCC in the middle Shire River catchment. Drought and floods in the western side of the river, therefore, have resulted into low crop yield. As a survival mechanism, people resort to cutting down of trees to earn income, causing forest degradation [58, 59]. This may, therefore, explain the concurrent low rainfall received against a sharp decline in forest areas between 2006 and 2011 (**Figures 2** and **3**). Results in this study agree with an earlier report for the upper Shire River catchment [60] indicating a direct link between poor rainfall (drought/floods) and cutting down of trees.

Rapid population growth is one of the drivers of LUCC in the western side of the middle Shire River earlier reported by [60, 61]. Population increase in the western part of the middle Shire River is mainly attributed to the influx of refugees fleeing the civil war from Mozambique from the 1990s. Population growth leads to urbanization, increase in cultivated land, and residential area [3, 8]. The high population density in Malawi with an estimated growth rate of 2.8% is putting increasing pressure on its natural resources, leading to expansion of farming on marginal lands and forests as well as encroachment into protected forest reserves/parks. Results in this study show a transition of land use from forest, shrubland, and grassland to cultivated land and buildup areas (**Tables 4** and **5**). These changes mainly occurred between 1989 and 2011 (**Figure 2** and **Table 4**) probably due to increasing anthropogenic pressure on natural forests. Results also show a drastic change in forest/grassland/shrubland between 1989 and 2011 in three out of the four districts (Balaka, Neno, and Ntcheu) in the western side of the middle River Shire. Large proportion of shrubland, grassland, and forestland (84%) in the western part of the river were converted to cultivated land, buildup areas, and/or bare land. This confirms earlier assertion that increasing population results into a decrease in forest area (**Figure 5**).

The rate of forest decline experienced by Malawi [61] and the Shire River catchment in particular [59], due to heavy dependency on wood for energy, is alarming. Most people around the middle Shire River catchment rely on firewood

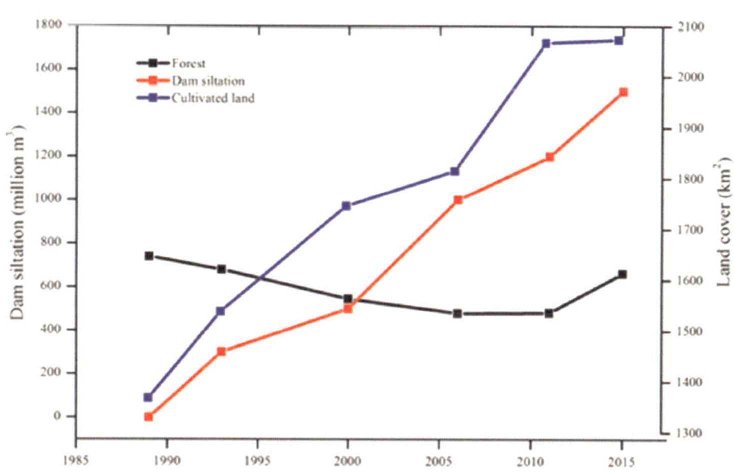

Figure 5.
Changes in forest, cultivated land in the catchment area, and siltation volume in the Nkula Dam from 1989 to 2015.

and charcoal for their daily living [58, 62, 63]. Malawi's forest cover loss is estimated at 2.6% per annum [64]. The middle Shire River catchment lost, on average, about 4.3% of its forest and shrubland annually between 1989 and 2011 (**Table 4**), suggesting a negative relationship between population increase and the decline in forest coverage (**Figures 4** and **5**). Results, nevertheless, showed a recovery in forest cover from 2011 to 2015 (**Tables 4** and **5**), likely attributed to interventions by the government of Malawi and nongovernmental organizations in strengthening natural resource management policies that started around 2008 up to date [5, 65].

Macroeconomic activities such as increase in manufacturing industries and other businesses which contribute to the growth of GDP often require large areas, which also contributed to the transition of forest/shrubland/grassland into buildup areas. Some of such economic activities include opening of new farms which also require clearing of forest areas (**Figures 4** and **5**).

National policies in the past have failed to effectively enforce ban of unabated harvesting of forest resources until recently with the introduction of community-based natural resource management groups and intervention of some nongovernmental organizations in afforestation programs. This may explain the increase in forest cover from 2011 to 2015 as earlier indicated (**Figure 2** and **Table 4**). Globally, large expanses of forests are being converted into bare land for domestic purposes and, principally, due to harvesting of timber [66]. In a study carried out between 1989 and 2002 in the upper section of the Shire River, [60] reported impacts of LUCC on the river's catchment hydrological regime which includes increase in soil erosion. It is reported that agricultural land increased by 18% between 1989 and 2002 [60]. In another LUCC assessment study for Likangala River catchment (a stream from Zomba Mountain which is also a source of several rivers draining into the eastern side of the middle Shire River), woodlands decreased from 135.3 km^2 in 1984 to 15.5 km^2 in 2013 [67]. These results agree with the present study confirming negative impacts of LUCC. Agriculture is the main source of employment to about 92% of the population in Malawi which lives in rural areas [61, 68]. Increase in agricultural activities leads to cultivated land expansion. Cash crops (e.g., tea, coffee, tobacco, and cotton), subsistence crops (e.g., maize and groundnuts), and animal rearing contribute to the increase in agricultural GDP. Results in the present study agree with a report for the region in which land use change (increase in farming activities) contributed to increase in GDP. Similar findings have also been reported correlating land use to increase in income [67]. The increase in cultivated land and artificial surfaces resulted into a decline in forest and shrubland (**Tables 4** and **5**).

Furthermore, the country loses about 1.7% of its GDP on average annually due to the combined effects of droughts and floods [69]. Heavy rains received during the 1989 season in the country (**Figure 3**) were associated with devastating floods that drastically affected the GDP due to crop failure and loss of property as well as human life in the same period but increased in the subsequent year (**Figure 4**). Although the devastating rainfall in the 1989 season played a role in influencing the GDP, other factors could also be at play due to the fact that drivers of economic growth are diverse and vary in the magnitude of influence. For example, in 1989, Malawi's economy was associated with high fuel prices due to the war in Mozambique. All fuel transportation routes from the Indian Ocean ports in Mozambique were blocked, and consequently, there was a collapse in commodity prices [68]. Poor sales of tobacco which is the country's major foreign exchange earner also affected the GDP in 1989 [68]. Increased GDP between 2005 and 2009 has been attributed to stabilization and enhanced income growth, which increased income per capita due to the new economic policies and a stable political environment in 2004 [68].

4.2 Consequences of forest decline

These study findings show a decline in forests and then an increase over the past 26 years (**Figures 2** and **5** and **Table 4**). Clearing of forests from the catchment of the middle Shire River has subjected the bare soil to erosion which finds its way into the Shire River downstream to the Nkula Dam as a sink. This, thus, may explain the heavy siltation at the Dam which has reduced the volume of water causing problems with normal generation of electricity (**Figures 4** and **5**). The volume of the Dam at Nkula Falls, which was 3 million m^3 at its construction in the 1980s, has recently dropped to nearly half of its original size due to massive siltation which consequently resulted in low production of hydroelectricity, now failing to meet the country's demand for power. Nkula B Hydroelectric Power Station is the main electricity generation plant in Malawi producing about 124 MW of electricity [70]. The electricity-providing company—the Electricity Supply Commission of Malawi (ESCOM)—is now implementing involuntary power load shedding programs resulting into national frequent blackouts. Consumers now resort to excessive use of firewood/charcoal in place of electricity for cooking and other domestic chores creating a heavy dependency on forest resources.

High soil losses in Ntcheu and Neno Districts could be due to increased population as a result of the refugees' long time settlement in these areas resulting into removal of forests. The expansion of cultivated land could thus be the cause for increased soil erosion and sediment transport downstream, which consequently accumulate in the Nkula Dam in the middle Shire River (**Figure 5**). These findings agree with a recent study [6] which confirmed that most of the sediments going into the Shire River and finally depositing at the Nkula Dam originate from the western side of the Shire River. Several studies elsewhere [20, 66] also report the same, linking increased population to deforestation and soil. Loss of forests coupled with agriculture are cause for rapid land use change resulting into increased soil erosion and siltation in the middle Shire River catchment [4, 6, 8] (**Figure 5**). Malawi, and the middle Shire River in particular, is therefore locked up in a cycle where anthropogenic activities in the river's catchment meant for a survival alternative to lack of electricity have become a cause for soil erosion and siltation in the river, consequently hampering the generation of the needed electricity.

5. Conclusions

Findings in this study show significant land use and land cover changes that have occurred in the middle Shire River catchment over the past 26 years which have also affected the Nkula Dam. Forestland and shrubland have declined, while cultivated land and artificial surfaces have increased in the area, and deforestation appears to be more pronounced in the western side of the middle Shire River. Severe siltation downstream in the Nkula Dam appears to be strongly linked to increased soil erosion as a result of land use and land cover change. Notable drivers for LUCC include rapid population growth and GDP, macroeconomic activities occurring especially in the western part of the river such as manufacturing industries, and poor national policies that have failed to effectively enforce ban of uncontrolled harvesting of forest resources.

To solve these problems, there is a need to review and amend weak policies that encourage noncompliance to regulations of managing forests. For example, all policies that may encourage or result in soil erosion such as river bank cultivation must be amended. Powers should be invested in local authorities to take part in protecting the environment and/or in planting trees, and the government should be able to provide seedlings for the operation. This should be done in a competition manner that the village which will perform well should be given some incentives. There is also need to increase fertilizer use so that land expansion for farming is curbed and yields are improved. In addition to that, population growth can be controlled through increase use of family planning. Encouraging children to go to school to avoid early marriages might also help to reduce poverty which will help to avoid cutting down of trees careless. Deliberate programs should be instituted by the government to curb further effects of climate variability such as droughts and floods. Such programs may include good agricultural practices that conserve soil and protect it from water erosion, discourage river bank cultivation, intensify afforestation programs, and ban the burning of charcoal. Findings in this study and the combination of methods used (application of GIS, remote sensing, and analysis of socioeconomic factors) can possibly be applied in areas where similar environmental problems have occurred. It is preferable to include a conclusion(s) section which will summarize the content of the book chapter.

Acknowledgements

We thank the State Key Laboratory of Estuarine and Coastal Research and Graduate School of East China Normal University (ECNU) for supporting this study. We also appreciate the valuable comments provided by Professor Christo C. P. Van der Westhuizen of North West University (South Africa), Professor Fang Shen of East China Normal University (China), Dr. Mavuto Tembo of Mzuzu University (Malawi), Ms. Lostina S. Chapola of Catholic University (Malawi), Mr. Tanazio Kwenda from the Department of Surveys (Malawi), Mr. Patrick Jambo from Forestry Department of Mzuzu University (Malawi), Mr. Samuel Limbu of the University of Dar es Salaam (Tanzania), Dr. Naziha Mokadem of North West University (South Africa), and the anonymous reviewers who helped us to polish this manuscript.

Conflict of interest

No potential conflict of interest was reported by the authors.

Author details

Maureen Kapute Mzuza[1,2*], Weiguo Zhang[3], Fanuel Kapute[2] and Xiaodao Wei[3]

1 North-West University, Potchefstroom, South Africa

2 Mzuzu University, Mzuzu, Malawi

3 State Key Laboratory of Estuarine and Coastal Research, Shanghai, China

*Address all correspondence to: maureenmzuza@yahoo.com

References

[1] Lambin EF, Geist HJ. Land use and land cover change: Local processes and global impacts. Environmental Sciences. 2006;**1**:1-8

[2] Sultan RM. The impacts of agricultural expansion and interest groups on deforestation: An optimal forest control model. International Journal of Agricultural Resources, Governance and Ecology. 2016;**12**(2): 137-154. DOI: 10.1504/IJARGE. 2016.076926

[3] Palamuleni LG. Land cover change and hydrological regimes in the Shire River Catchment, Malawi [doctoral dissertation]. Johannesburg: University of Johannesburg; 2009

[4] Wiyo KA, Fiwa L, Mwase W. Solving deforestation, protecting and managing key water catchments in Malawi using smart public and private partnerships. Journal of Sustainable Development. 2015;**8**(8):251. DOI: 10.5539/jsd. v8n8p251

[5] Mzuza MK, Chapola L, Kapute F, Chikopa I, Gondwe J. Analysis of the impact of aquatic weeds in the Shire River on generation of electricity in Malawi: A case of Nkula Falls hydro-electric power station in Mwanza District, Southern Malawi. International Journal of Geosciences. 2015;**6**(06):1-8

[6] Mzuza MK, Weiguo Z, Chapola LS, Tembo M, Kapute F. Determining sources of sediments at Nkula Dam in the Middle Shire River, Malawi, using mineral magnetic approach. Journal of African Earth Sciences. 2017;**126**:23-32. DOI: 10.1016/j.jafrearsci.2016.11.023

[7] Dulanya Z. A review of the geomorphotectonic evolution of the south Malawi rift. Journal of African Earth Sciences. 2017;**129**:728-738. DOI: 10.1016/j.jafrearsci.2017.02.016

[8] Bell AR, Ward PS, Mapemba L, Nyirenda Z, Msukwa W, Kenamu E. Smart subsidies for catchment conservation in Malawi. Scientific Data. 2018;**5**:180113

[9] Musakwa W, Wang S. Landscape change and its drivers: A Southern African perspective. Current Opinion in Environmental Sustainability. 2018;**33**: 80-86. DOI: 10.1016/j.cosust. 2018.05.001

[10] Kaunda CS, Mtalo F. Impacts of environmental degradation and climate change on electricity generation in Malawi. International Journal of Energy & Environment. 2013;**4**(3):481-496

[11] Xu HQ, Wang XQ, Xiao GR. A remote sensing and GIS integrated study on urbanization with its impact on arable lands: Fuqing City, Fujian Province, China. Land Degradation & Development. 2000;**11**(4):301-314. DOI: 10.1002/1099-145X(200007/08) 11:4<301:AID-LDR392>3.0.CO;2-N

[12] Meshesha TW, Tripathi SK, Khare D. Analyses of land use and land cover change dynamics using GIS and remote sensing during 1984 and 2015 in the Beressa Watershed Northern Central Highland of Ethiopia. Modeling Earth Systems and Environment. 2016;**2**(4): 168

[13] Worku T, Khare D, Tripathi SK. Modeling runoff–sediment response to land use/land cover changes using integrated GIS and SWAT model in the Beressa watershed. Environmental Earth Sciences. 2017;**76**(16):550

[14] Yang K, Lu C. Evaluation of land-use change effects on runoff and soil erosion of a hilly basin—The Yanhe River in the Chinese Loess Plateau. Land Degradation & Development. 2018; **29**(4):1211-1221. DOI: 10.1002/ldr.2873

[15] Eaton D. The Economics of Soil Erosion: A Model of Farm Decision-making. Londres: International Institute for Environment and Development; 1996. pp. 1-48

[16] Greenough G, McGeehin M, Bernard SM, Trtanj J, Riad J, Engelberg D. The potential impacts of climate variability and change on health impacts of extreme weather events in the United States. Environmental Health Perspectives. 2001;**109**(Suppl 2):191. DOI: 10.1289/ehp.109-1240666

[17] Ouellet C, Saint-Laurent D, Normand F. Flood events and flood risk assessment in relation to climate and land-use changes: Saint-François River, southern Québec, Canada. Hydrological Sciences Journal. 2012;**57**(2):313-325. DOI: 10.1080/02626667.2011.645475

[18] Alexakis DD, Grillakis MG, Koutroulis AG, Agapiou A, Themistocleous K, Tsanis IK, et al. GIS and remote sensing techniques for the assessment of land use change impact on flood hydrology: The case study of Yialias basin in Cyprus. Natural Hazards and Earth System Sciences. 2014;**14**(2): 413-426. DOI: 10.5194/nhessd-1-4833-2013

[19] Davies GM, Pollard L, Mwenda MD. Perceptions of land-degradation, forest restoration and fire management: A case study from Malawi. Land Degradation & Development. 2010;**21**(6):546-556. DOI: 10.1002/ldr.995

[20] Mohawesh Y, Taimeh A, Ziadat F. Effects of land use changes and soil conservation intervention on soil properties as indicators for land degradation under a Mediterranean climate. Solid Earth. 2015;**6**(3):857-868. DOI: 10.5194/se-6-857-2015

[21] Bishop J. The Cost of Soil Erosion in Malawi. Washington, DC: Draft report for World Bank; 1990

[22] Barbier EB, Burgess J. Agricultural pricing and environmental degradation. In: Background Paper for World Bank. World Development Report. 1992

[23] Chavula G, Brezonik P, Bauer M. Land use and land cover change (LULC) in the Lake Malawi Drainage Basin, 1982-2005. International Journal of Geosciences. 2011;**2**(02):172. DOI: 10.4236/ijg.2011.22018

[24] Malawi Government. Land Husbandry Department. Chitedze Agriculture Research Station Official Report; Lilongwe, Malawi. 2015

[25] Coulibaly JY, Mbow C, Sileshi GW, Beedy T, Kundhlande G, Musau J. Mapping vulnerability to climate change in Malawi: Spatial and social differentiation in the Shire River Basin. American Journal of Climate Change. 2015;**4**(03):282. DOI: 10.4236/ajcc.2015.43023

[26] Schäfer MP, Dietrich O, Mbilinyi B. Streamflow and lake water level changes and their attributed causes in Eastern and Southern Africa: State of the art review. International Journal of Water Resources Development. 2016;**32**(6): 853-880. DOI: 10.1080/07900627.2015.1091289

[27] Gessesse B, Bewket W, Bräuning A. Model-based characterization and monitoring of runoff and soil erosion in response to land use/land cover changes in the Modjo watershed, Ethiopia. Land Degradation & Development. 2015; **26**(7):711-724. DOI: 10.1002/ldr.2276

[28] Ligonja PJ, Shrestha RP. Soil erosion assessment in kondoa eroded area in Tanzania using universal soil loss equation, geographic information systems and socioeconomic approach. Land Degradation & Development. 2015;**26**(4):367-379. DOI: 10.1002/ldr.2215

[29] Zhang F, Tiyip T, Feng ZD, Kung HT, Johnson VC, Ding JL, et al. Spatio-temporal patterns of land use/cover changes over the past 20 years in the middle reaches of the Tarim River, Xinjiang, China. Land Degradation & Development. 2015;**26**(3):284-299. DOI: 10.1002/ldr.2206

[30] Tsendbazar N, Herold M, Lesiv M, Fritz S. Copernicus Global Land Operations—Vegetation and Energy "CGLOPS-1". 2018

[31] Kapetsky JM, Aguilar-Manjarrez J. Geographic information systems, remote sensing and mapping for the development and management of marine aquaculture. In: Food & Agriculture Org. Technical Paper No. 458. 2007. p. 125

[32] Simic A, Fernandes R, Wang S. Assessing the impact of leaf area index on evapotranspiration and groundwater recharge across a shallow water region for diverse land cover and soil properties. Journal of Water Resource and Hydraulic Engineering. 2014;**3**:60-73

[33] Kaunda CS. Energy situation, potential and application status of small-scale hydropower systems in Malawi. Renewable and Sustainable Energy Reviews. 2013;**26**:1-9. DOI: 10.1016/j.rser.2013.05.034

[34] Gamula GE, Hui L, Peng W. Development of renewable energy technologies in Malawi. International Journal of Renewable Energy Technology Research. 2013;**2**(2): 44-52

[35] Shela ON. Naturalisation of Lake Malawi levels and Shire River flows: Challenges of water resources research and sustainable utilisation of the Lake Malawi-Shire River system. In: Water Net Symposium: Sustainable Use of Water Resources. 2000. pp. 1-12

[36] Malawi Meteorological Services. Climate of Malawi; 2013; Malawi. Lilongwe: 2013

[37] Morel SW. Chemical mineralogy and geothermometry of the middle Shire granulites, Malawi. Journal of African Earth Sciences (and the Middle East). 1989;**9**(1):169-178. DOI: 10.1016/0899-5362(89)90018-3

[38] Jiménez A, Vilchez F, González O, Flores S. Analysis of the land use and cover changes in the metropolitan area of Tepic-Xalisco (1973–2015) through landsat images. Sustainability. 2018; **10**(6):1860. DOI: 10.3390/su10061860

[39] Pettorelli N, Vik JO, Mysterud A, Gaillard JM, Tucker CJ, Stenseth NC. Using the satellite-derived NDVI to assess ecological responses to environmental change. Trends in Ecology & Evolution. 2005;**20**(9): 503-510. DOI: 10.1016/j.tree.2005.05.011

[40] Foody GM, Campbell NA, Trodd NM, Wood TF. Derivation and applications of probabilistic measures of class membership from the maximum-likelihood classification. Photogrammetric Engineering and Remote Sensing. 1992;**58**(9):1335-1341

[41] Bolstad P, Lillesand TM. Rapid maximum likelihood classification. Photogrammetric Engineering and Remote Sensing. 1991;**57**(1):67-74

[42] Gupta S, Islam S, Hasan MM. Analysis of impervious land-cover expansion using remote sensing and GIS: A case study of Sylhet sadar upazila. Applied Geography. 2018;**98**:156-165. DOI: 10.1016/j.apgeog.2018.07.012

[43] Halefom A, Teshome A, Sisay E, Ahmad I. Dynamics of land use and land cover change using remote sensing and GIS: A case study of Debre Tabor Town,

South Gondar, Ethiopia. Journal of Geographic Information System. 2018; **10**(02):165-174. DOI: 10.4236/ jgis.2018.102008

[44] Ornetsmüller C, Heinimann A, Verburg PH. Operationalizing a land systems classification for Laos. Landscape and Urban Planning. 2018; **169**:229-240. DOI: 10.1016/j. landurbplan.2017.09.018

[45] Payne C, Panda S, Prakash A. Remote sensing of river erosion on the Colville river, North Slope Alaska. Remote Sensing. 2018;**10**(3):397. DOI: 10.3390/rs10030397

[46] Jensen J. Residual maximum likelihood estimation of (co)variance components in multivariate mixed linear models using average information. Journal of the Indian Society of Agricultural Statistics. 1997; **49**:215-236

[47] Anderson JR. A Land Use and Land Cover Classification System for Use with Remote Sensor Data. US: Government Printing Office; 1976;**964**

[48] Weng Q. Land use change analysis in the Zhujiang Delta of China using satellite remote sensing, GIS and stochastic modelling. Journal of Environmental Management. 2002; **64**(3):273-284. DOI: 10.1006/ jema.2001.0509

[49] Foody GM. Status of land cover classification accuracy assessment. Remote Sensing of Environment. 2002; **80**(1):185-201. DOI: 10.1016/ S0034-4257(01)00295-4

[50] Foody GM. Assessing the accuracy of land cover change with imperfect ground reference data. Remote Sensing of Environment. 2010;**114**(10):2271-2285. DOI: 10.1016/j.rse.2010.05.003

[51] Stehman SV, Czaplewski RL. Design and analysis for thematic map accuracy assessment: Fundamental principles. Remote Sensing of Environment. 1998; **64**(3):331-344. DOI: 10.1016/ S0034-4257(98)00010-8

[52] Pauw K, Thurlow J, Bachu M, Van Seventer DE. The economic costs of extreme weather events: A hydrometeorological CGE analysis for Malawi. Environment and Development Economics. 2011;**16**(2):177-198. DOI: 10.1017/S1355770X10000471

[53] NSO (National statistical Office). Malawi Statistics 2012-2017. Zomba: National Statistical Office. 2017

[54] Nicholson SE, Klotter D, Chavula G. A detailed rainfall climatology for Malawi, Southern Africa. International Journal of Climatology. 2014;**34**(2): 315-325. DOI: 10.1002/joc.3687

[55] Kumbuyo CP, Yasuda H, Kitamura Y, Shimizu K. Fluctuation of rainfall time series in Malawi: An analysis of selected areas. Geofizika. 2014;**31**(1): 13-28. DOI: 15233/gfz.2014.31.1

[56] Clay E, Bohn L, de Armas EB, Kabambe S, Tchale H. Malawi and Southern Africa: Climatic variability and economic performance. Disaster Management Facility. Washington DC: World Bank Working Paper Series (No. 7); 2003

[57] Kambewa P. Charcoal—The reality: A study of charcoal consumption, trade, and production in Malawi. Iied; 2007

[58] Hudak AT, Wessman CA. Deforestation in Mwanza District, Malawi, from 1981 to 1992, as determined from Landsat MSS imagery. Applied Geography. 2000;**20**(2): 155-175. DOI: 10.1016/S0143-6228(00) 00002-3

[59] Nanthambwe S. Policy Sector Review for incorporating Sustainable

Land Management in the Shire River Basin and development of an institutional framework for sustainable land management. Final Report, Government of Malawi, Environmental Affairs Department, Ministry of Environment and Climate Change Management; Lilongwe; 2013

[60] Palamuleni LG, Ndomba PM, Annegarn HJ. Evaluating land cover change and its impact on hydrological regime in Upper Shire river catchment, Malawi. Regional Environmental Change. 2011;**11**(4):845-855

[61] Tobin RJ, Knausenberger WI. Dilemmas of development: Burley tobacco, the environment and economic growth in Malawi. Journal of Southern African Studies. 1998;**24**(2):405-424. DOI: 10.1080/03057079808708582

[62] Mlotha MJ. Remote sensing and GIS linked to socio-analysis for land cover change assessment. In: Geoscience and Remote Sensing Symposium, 2001. IGARSS'01. IEEE 2001 International. IEEE; 2001. pp. 459-461. DOI: 10.1109/IGARSS.2001.976189

[63] Taulo JL, Gondwe KJ, Sebitosi AB. Energy supply in Malawi: Options and issues. Journal of Energy in Southern Africa. 2015;**26**(2):19-32

[64] Mwanakatwe P. African Economic Outlook. In: Malawi 2015. AfDB, OECD: UNDP; 2015

[65] Zulu LC. The forbidden fuel: Charcoal, urban wood fuel demand and supply dynamics, community forest management and wood fuel policy in Malawi. Energy Policy. 2010;**38**(7):3717-3730. DOI: 10.1016/j.enpol.2010.02.050

[66] Lambin EF, Turner BL, Geist HJ, Agbola SB, Angelsen A, Bruce JW, et al. The causes of land-use and land-cover change: Moving beyond the myths. Global Environmental Change. 2001;**11**(4):261-269. DOI: 10.1016/S0959-3780(01)00007-3

[67] Pullanikkatil D, Palamuleni L, Ruhiiga T. Assessment of land use change in Likangala River catchment, Malawi: A remote sensing and DPSIR approach. Applied Geography. 2016;**71**:9-23. DOI: 10.1016/j.apgeog.2016.04.005

[68] Ngwira AR, Aune JB, Mkwinda S. On-farm evaluation of yield and economic benefit of short term maize legume intercropping systems under conservation agriculture in Malawi. Field Crops Research. 2012;**132**:149-157. DOI: 10.1016/j.fcr.2011.12.014

[69] Pauw K, Thurlow J, Van Seventer D. Droughts and Floods in Malawi. No. 962. Assessing the Economy Wide Effects. Malawi: Discussion Paper, International Food Policy Research Institute (IFPRI); 2010

[70] ESCOM (Electricity Supply commission of Malawi). 2018. Available from: http://www.escom.mw/waterlevels-energysituation-malawi.php [Accessed: 2018-08-14]

[71] FAO (The Food and Agriculture Organization). Global forest resources assessment 2015. Country Report, Malawi, 2015. Available from: http://www.fao.org/3/a-az265e.pdf [Accessed: 2018-11-20]

Chapter 4

Advanced Methods for Spatial Analysis of Bioaerosol Long-Range Transport Processes

Daniel A. Pickersgill, Hartmut Müller and Viviane R. Després

Abstract

Research on bioaerosol is still in its infancy. The dynamics and, therefore, the effects on atmospheric processes and the biosphere are often underestimated, or have not yet been sufficiently investigated. Atmospheric models such as FLEXPART and HYSPLIT enable researchers to simulate the transport of particles in the atmosphere and provide information on where air-parcels originate from. In the following, we present two methods for combining results of these models with spatial information, e.g., about vegetation. The first method shows how spatial CORINE land cover distribution can be analyzed within the boundaries of HYSPLIT trajectories. In a second method, FLEXPART simulations are used in combination with COSMO rain data and tree maps to generate maps that indicate the potential origin of bioaerosol for selected periods of time.

Keywords: bioaerosol, wind trajectories, FLEXPART, GIS, ArcPy

1. Introduction

Bioaerosol, more precisely primary biological aerosol particles (PBAP), are particles of biological origin, smaller than 100 μm, that are released into the atmosphere. These include viruses, bacteria, pollen and fungal spores, small and cell fragments, and excrements from organisms [1]. The diameter of PBAP ranges from a few nanometers, such as cell fragments, proteins and viruses, to the upper size boundary seen for many plant pollen [2].

PBAPs emitted into the atmosphere are subject to many physical factors resulting in a considerable influence on the atmospheric residence time. Many meteorological factors such as wind speed, wind direction, convection, temperature and relative humidity influence the residence time which can result in PBAP covering long distances in the atmosphere. In addition, the residence time depends strongly on the aerodynamic diameter of the particles [3]. Removal from the atmosphere can happen by dry and wet deposition. For dry deposition, the PBAPs are removed by sedimentation, while for wet deposition they are washed out by precipitation. PBAP have a direct influence on humans when they act as allergens or pathogens [4, 5]. As an increasing proportion of the population suffers from allergies, the importance of accurate pollen prediction increases. Predictions are currently generated by the Deutscher Wetterdienst DWD (German Weather Service), which issues warnings

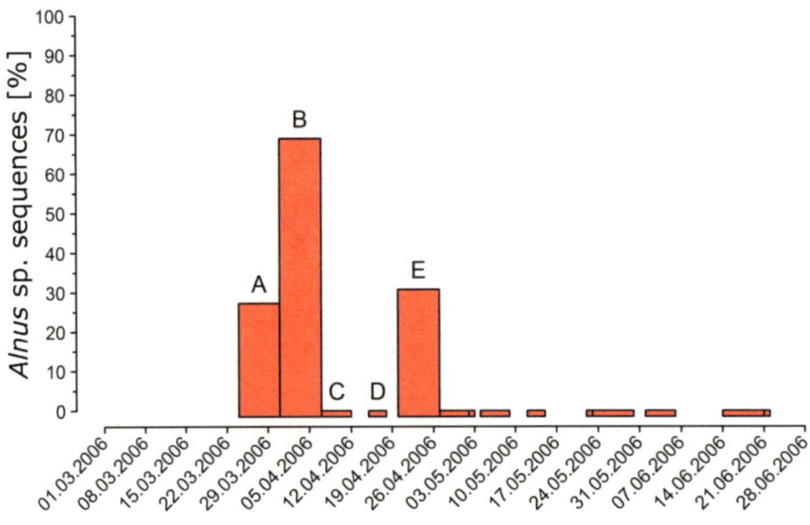

Figure 1.
Proportion of Alnus sp. (alder) DNA sequences in the total number of isolated plant sequences collected between March and June 2006 in Mainz, Germany (49°59'31.36"N 8°14'15.22"E). In addition to the percentage (y-axis) the time period of the individual air samples is visualized by the column width (x-axis).

for very large areas if a certain pollen concentration in the air is present. The pollen concentration is determined by the German Pollen Information Service Foundation (http://www.pollenstiftung.de) through microscopic analysis. The DWD processes these data using weather models and produces warning maps for pollen. Due to the long processing path and the small number of pollen collecting stations, large inaccuracies occur. A more precise knowledge of the emission and transport processes could lead to considerably improved pollen predictions and thus help patients to be better prepared.

The analysis of PBAP, especially concerning their origin, poses special challenges for science. Atmospheric transport is a highly complex issue. The composition of emissions, strongly depend on the biotope, which is highly variable in space, and in time, dependent on the season. **Figure 1** shows an example of the proportion of *Alnus* sp. (alder) DNA sequences in the total number of isolated plant sequences found on analyzed weekly air sample filters in spring 2006. As can be seen, the proportion decreases from sample B to C from ~70 to 0% and increases from sample D to E from 0 to ~30%. These results raise the question whether such high variances in such a short time frame can be explained by air movement.

In the following sections we will describe methods to identify potential areas of origin of PBAP starting from a firmly defined sampling location and using Lagrangian back-trajectory transport model. By using two-dimensional raster data containing information on potential emission sources we will show methods to identify correlations with observed PBAP data by modelling the potential transport processes of the particles in the atmosphere.

2. Data

2.1 PBAP data

The aerosols were collected on glass fiber filters for over a year. The data collection station was established at about 20 m height above ground, on the roof of the old Max Planck Institute for chemistry, located at the campus of the Johannes Gutenberg University Mainz (49°59′31.36"N, 8°14′15.22"E). An air volume of 0.3 m^3 min^{-1} was filtered by a high-volume sampler. The operating time was between 1 and 7 days, which was equivalent to an air volume of about 430–3000 m^3. The DNA was extracted from the air filter samples, the plant DNA isolated and taxonomically identified. The previously unpublished data were kindly provided by Isabell Müller-Germann [6]. In pollen flight periods a large percentage of plant bioaerosol are formed by pollen. In the following, therefore, the simplified term pollen is used, even though other plant particles may have been measured. An exact description of the methodology can be found in Fröhlich-Nowoisky et al. [7], with the modification that plant-specific primers were used.

2.2 HYSPLIT

HYSPLIT (HYbrid Single-Particle Lagrangian Integrated Trajectory) [8] has been used for about 30 years by the atmospheric science community to calculate atmospheric transport, dispersion, chemical transformation and deposition. A major task of the system is the calculation of backward trajectories to determine the origin of air masses. This makes it possible to establish cause-effect relationships. Forward trajectories, on the other hand, are used to predict the propagation of, e.g., volcanic ash or radioactive particles. HYSPLIT uses a hybrid calculation approach and uses both Lagrangian and Eulerian methods [9]. By using the READY system (http://ready.arl.noaa.gov/index.php), HYSPLIT calculations can even be performed online. The meteorological basis of the calculations is the globally available "Global Data Assimilation System" (GDAS) provided by the National Oceanic and Atmospheric Administration NOAA with an area resolution of 1 × 1 and a vertical subdivision into 23 layers reaching a height of 26.5 km above ground. The results of a back trajectory calculation are single line features obtained for each start time of the calculation. The line feature contains the four-dimensional information (place and time) on the origins of an air-parcel that was measured at a defined point in time t0. The result of the calculation can be generated in zipped Keyhole Markup Language (.kmz) or Shapefile (.shp) format, which easies further processing in Geographic Information Systems.

2.3 FLEXPART

FLEXible PARTicle dispersion model (FLEXPART) [10] is a Lagrangian particle dispersion model for calculating the propagation of air masses over long distances. The results are obtained in the multidimensional Network Common Data Form (NetCDF) format which is often used in meteorology [11]. The result files contain spatial information for five air layers and two particle types (tracer). One particle type has an atmospheric half-life of 12 hours, which approximately is equal to the average atmospheric retention time for PBAP of pollen size, when considering dry and wet deposition. For comparison reasons, no half-life parameter was set for the second particle type. It thus represents a so-called air tracer. For every single day of a 10 weeks' period a simulation was performed with a spatial resolution of 10 × 10 km.

The FLEXPART model calculation is based on four-dimensional meteorological raster data sets, which define the level of resolution for the results. The DWD used analysis data from the COSMO-EU model [12] for this purpose (http://www.dwd.de). The data resolution is 0.0625° (~7 km) in horizontal direction. In vertical direction, the raster cells are ordered into 40 layers reaching a height of 22.5 km above ground. The raster grid is rotated against the north direction, with the North Pole at 40° latitude and −170° longitude.

In addition, rain areas (see Section 3.2.3) were extracted from the COSMO-EU model. These areas contain information about location, quantity and type of rain (convective or scalar), given in the form of total hourly rainfall accumulated to daily sums in mm.

2.4 Tree species maps for European forests

The "Tree species maps for European forests" published by the European Forest Institute [13] were used as a data source for potential pollen emission sites. These maps were generated using statistical methods such as logistic regression and Kriging and differ according to region and national forest inventory methodology. The resulting data sets represent the only European-wide mapping of tree species to date. The data for each recorded tree species is provided in the form of a GeoTiff raster layer with a resolution of 1 × 1 km per grid cell. The numerical value of each grid cell is the percentage coverage of the cell with the respective tree species.

The tree species maps are available in the spatial reference system ETRS89/ ETRS-LAEA (EPSG:3035). For the processing discussed here they were transformed into the geodetic reference system WGS84.

2.5 CORINE land cover

Another used data source is the land use map published by the European Environment Agency's "CORINE Land Cover" project [14]. Updates to the CORINE Land Cover (CLC) inventory, dating from1985, are available for 2000, 2006, 2012, and 2018. The 2006 version best fit the time frame of the PBAP data and was therefore used. The map shows the type of land use for the participating 39 EU countries in 100 × 100 m grid cell resolution, subdivided into 44 different land-use classes. The map is available in the spatial reference system ETRS89/ETRS-LAEA (EPSG:3035) and, therefore, was also transformed into WGS84 for processing.

3. Methodology

3.1 Land use analyses based on trajectories (HYSPLIT)

Trajectories give information about height, direction and residence time of an air parcel. The used backward trajectories show which ground areas have been over flown before an air parcel reached the measuring point in Mainz. The area composition has a decisive influence on the composition of the collected bioaerosol.

The trajectories calculated with HYSLPIT are only available as line-features and do not provide information that could serve as a basis for area-related calculation. To solve this problem area-buffers were created for the line features. In a second step, these buffers were used to calculate the composition of the underlying CORINE land use areas by polygon overlay (**Figure 2**). The resulting area clips of the data set could then be used to statistically evaluate the pollen values.

3.2 Potential maps of pollen origin

Both FLEXPART and HYSPLIT simulations generate information on the residence time of air parcels arriving at specific points of measurement. However, FLEXPART performs not only line calculations for individual points in time of the residence time as HYSPLIT trajectories do, but can generate grids of air parcel residence time for entire time periods. Therefore, in addition to location and time, one

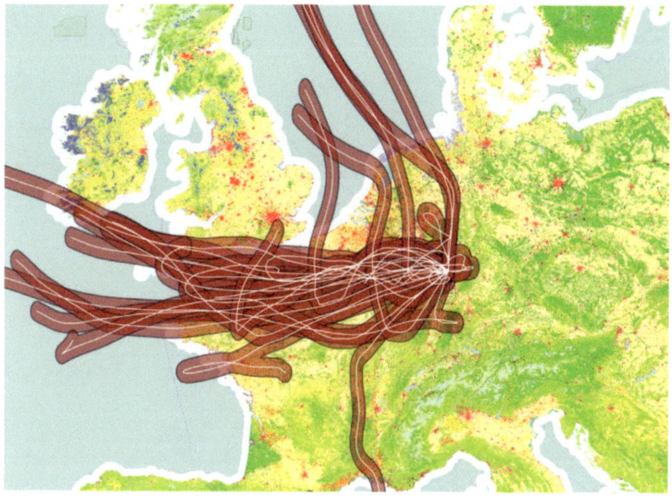

Figure 2.
HYSPLIT trajectories (lines) and their buffers superimposed on the CORINE land cover map.

gets the area-related information on how long air parcels have resided in defined areas. Thus, different weights are assigned to areas dependent on the influence on the air parcels that reach the point of measurement.

As discussed in Section 2, the FLEXPART day simulation results for the study periods are available in multidimensional NetCDF format, the format frequently used in meteorology. Due to the high number of possible two-dimensional raster representations the interpretation of the results in view of the transport of air parcels to the point of measurement is very time-consuming. It therefore makes sense to automate the processing steps. Processing steps to be automated include the export of information from NetCDF files into a raster format suited for further processing in a GIS, the subsequent processing of the raster grids including removing days of rain, and the concatenation of this information with tree cover gradients (see Section 2). The aim is to create maps which, by combining the residence time of air parcels with the tree population, lead to new insights into the potential origin of bioaerosol.

The FLEXPART results (see **Figure 3**) are extracted by using ArcPy, a Python library that provides tools for the analysis and conversion of geographical data [15]. Using tools from the "Multidimension toolbox," the spatial information for all required combinations of time period, height of the air parcels and tracer is extracted and saved as a raster file in GeoTiff format. Defining the required dimensions for NetCDF files with massive content is often a complex task. Therefore, an ArcGIS tool was developed that considerably simplifies the handling of dimensions for NetCDF files and data export [16]. The generated raster layers contain the residence time of the air parcels per day in a resolution of 10 × 10 km.

The individual layers created for each time step must be further processed. For the following steps only the two bottom air mass layers are considered. It is assumed that the so-called atmospheric boundary layer air is a homogeneously mixed in respect to the contained particles [17]. Solar radiation causes the boundary layer to build up during the course of the day. Depending on the conditions, the height of the boundary layer varies, but should normally be higher than 500 m above ground. Consequently, the two bottom air layers (height above ground 0–100 m, and height above ground 100–500 m) are merged into on single layer.

The tree species coverage data are available in a different projection system and at a higher resolution (1 × 1 km). Therefore, the FLEXPART grids are re-projected

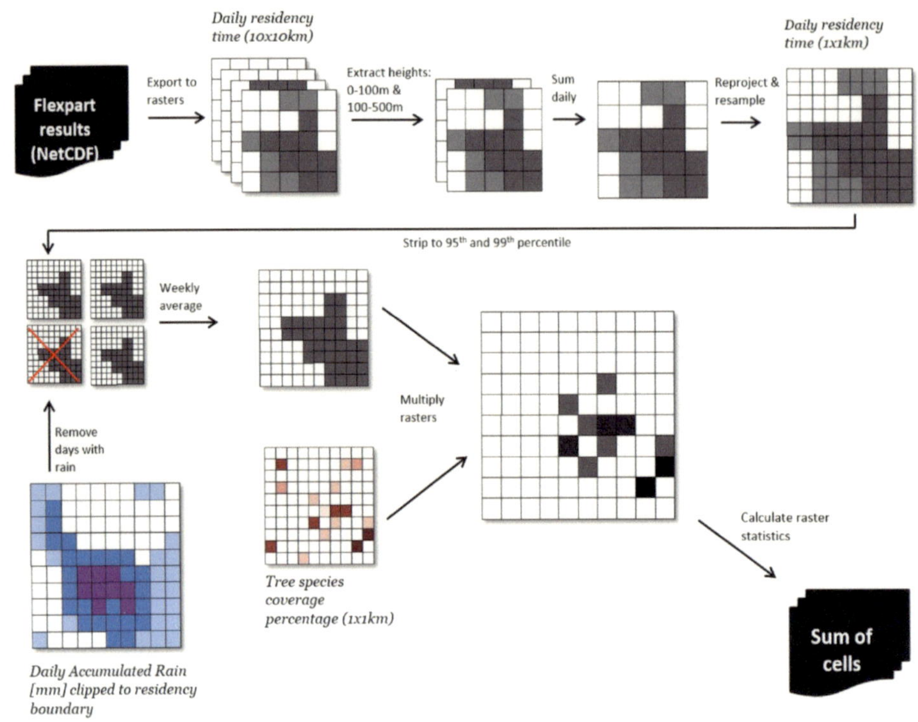

Figure 3.
Schematic representation of the most relevant processing steps.

and resampled according to the daily residency time. Re-projection eliminates pixel inconsistencies between different projection systems and thus prevents errors in raster operations such as multiplication. Although the resampling does not improve the FLEXPART data resolution, the resampled data can be used for subsequent analysis with tree species coverage data.

The temporal resolution of the pollen data analysis is usually 7 days; therefore, the daily residence times must also be averaged. Before doing so, days with rain are identified and excluded from the means formation to eliminate the influence of such days on the distribution of pollen in the air due to wet deposition [18].

For this purpose, hour values of rain fields which are available in grid format in the COSMO model are used (see Section 2). The data is provided in COSMO's native model grid and must first be rotated spatially to match the spatial reference system used in this study. In order to avoid the processing of uninvolved areas the rain grids are clipped to the spatial extent of the FLEXPART data sets for the considered periods, day sums are formed for the remaining rain cells. If the value of a day sum exceeds 2 mm per day, the air residency time for this day is removed, i.e., not considered in the week average value calculation.

The weekly average is the mean residence times of the remaining days. **Figure 4** shows examples of results for two different weeks. The figure illustrates the difference of these 2 weeks; as already mentioned, only dry days were taken into account.

A first attempt was made to derive land cover statistics for the areas under the air parcels, in analogy to the trajectory method. However, due to the large spatial extent following from the calculation of week average values, the statistics generated were not significant. Therefore, it was not possible to confine areas of potential origin of pollen with sufficient probability. A further attempt to produce significant statistics was made by extracting only areas with 99 and 95 percentile values, but with the persistent problem that all extracted areas had the same weight.

In a second approach, the weekly residence times of the air parcels was multiplied by the percentage of individual tree species coverage such as *Alnus* sp. or *Betula* sp. This approach yields high values for areas characterized by long residency periods and high coverage rates, low values for areas with short residency periods

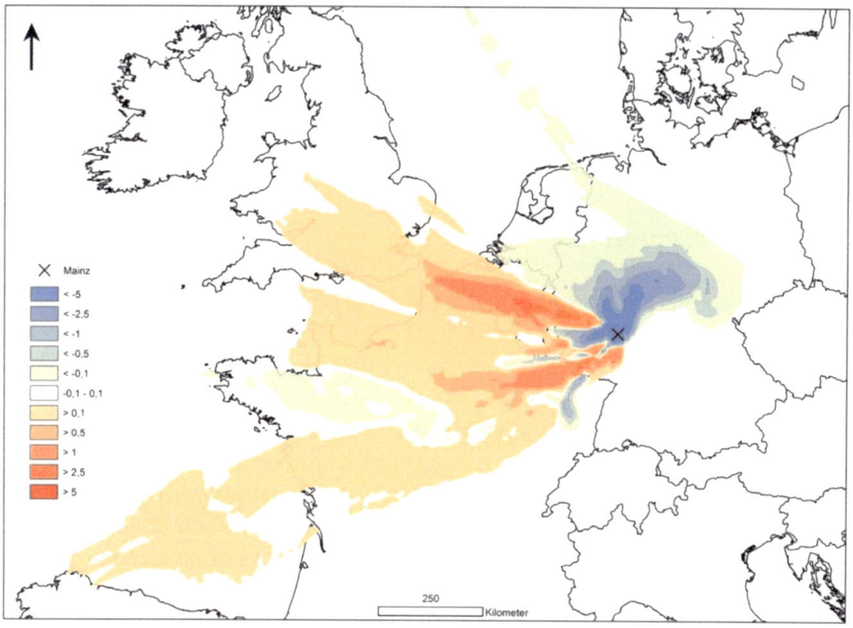

Figure 4.
Comparison of 2 weeks. The difference of residency time in hours per week is shown. A predominance of the first week is shown with a red color, the second in blue color.

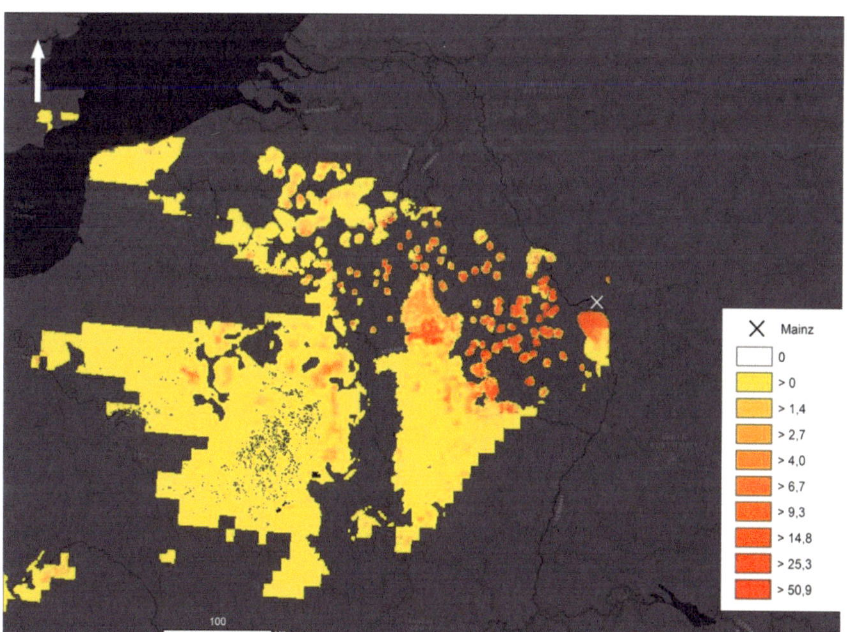

Figure 5.
Potential map. Red: high potential, yellow: low potential, transparent: no potential, because of no tree population or no residency time of an air parcel. Values result from multiplication of tree species coverage in percent and average residency time of air in hours per week.

and low coverage rates. This approach not only suppresses areas with a short residency time of an air parcel, which does not significantly affect the air composition at the point of measurement, but also areas with a long residency time of an air parcel, in areas without population of the respective tree species. The resulting potential maps show tree species populations that could be potential sources of detected pollen at the point of measurement in a given week (see **Figure 5**).

The resulting weekly potential maps can now be visually interpreted. Due to the large number of possible potential grids, however, it is useful to determine a numerical value for the purpose of automated evaluation. Summing up all grid cell values should result in high values for weeks with a high pollen volume.

4. Results and discussion

The described methods open up new possibilities for the analysis of long-distance transport processes in bioaerosol research. The research question was to establish a quantifiable link between temporally dynamic PBAP data, which are sampled at a spatially static point, and the spatially dynamic but temporally constant GIS raster maps. Both the temporally and the spatially resolved dispersion models function as connecting elements. The observation of near-ground air movements makes it possible to strongly limit the potential emission areas. Flexible calculations with high temporal resolution for each sampling period can be carried out on the basis of raster maps.

In contrast to HYSPLIT trajectories, the residence times of air parcels generated by FLEXPART is not specified at individual points in time, but for entire periods—instead of line features, grids are generated. This makes it possible to assign lower weights to areas with a short residence time of air parcel. In addition, FLEXPART takes the air movements in the near-ground layers of the atmosphere into account in a more realistic manner. Another advantage over HYSPLIT is the higher spatial and temporal resolution of the underlying data set (GDAS vs. COSMO-EU).

By multiplying by the percentage of tree species coverage, potential emission areas can be emphasized or omitted, depending on the occurrence of observed tree species. Consequently, only tree species populations that could be potential sources for pollen deposition within a given week are present in the resulting potential maps. Another advantage is the consideration of rain; it can be assumed that convective and continuous scalar showers efficiently remove aerosols at the point of origin.

The used greatly simplified filtering of rain events is still a source of uncertainty. If the amount of rain on a day exceeded the threshold value, the residence time of all air parcels for that day is not taken into account when calculating the average week value. A fixed threshold value is used and the temporal distribution and the type of precipitation are not considered. A short period of strong convective precipitation with a total of more than 2 mm is equal to a scalar precipitation spread over several hours. In addition, the spatial distribution of precipitation is not taken into account. Although only rain events within significant air parcels are considered, the same weight is assigned to all locations within these areas. Given the large spatial extent of the air masses, locations with substantially different distances to the point of measurement are equally weighted. Likewise, if a precipitation event is locally limited, all non-influenced air parcels for this day are removed.

5. Conclusions and outlook

Atmospheric convection and transport processes are highly complex issues for which many new insights have been gained in recent years. Hence, highly complex interactions, such as the transport of bioaerosol in the atmosphere including its physical and biological effects on different areas of the ecosystem, can be studied and evaluated.

The presented methods allow the confinement of potential emission sources of bioaerosol and could aid in assessing the contribution of long-range-transport to the locally measured bioaerosol. This, in a simpler first approach can be used to calculate the relative contribution of different potential emission areas, such as the presented land cover maps. This allows the search for correlations between a specific land-type with the occurrence of specific PBAP. Moreover, by combining the residence time of air-masses with rasterized coverage data, such as the discussed tree species maps, more accurate predictions on the potential contribution of certain areas can be made. This allows the analysis of the atmospheric transport of specific species or groups of species.

However, the developed methods are limited to the extent that it is not yet possible to adequately describe or simulate real conditions in several respects. A clear improvement would result from a trajectory-level consideration of rain events, which effectively remove bioaerosol from the atmosphere. For this purpose, it would be necessary to investigate each air parcel at all time-points to assess whether it is located in a rain event. If an air parcel crosses a rain event, the trajectory could be removed and thus not used for the calculation of the total residency time.

In all respects, the temporal and/or spatial resolution of the PBAP data, meteorological data, therefore the accuracy of model predictions, and the rasterized potential emission source maps, are key for a successful application of the methods. Constantly improving methods of bioaerosol monitoring, more detailed computational models, better computational performances and more accurate geographical data, will lead to more accurate analyses of PBAP transport processes in the atmosphere. Predicting bioaerosol composition and concentrations with a high spatial and temporal resolution would provide an excellent basis for assessing the impacts of bioaerosol on humans, the ecosystems and the climate and ultimately allow the implementation of accurate early-warning systems to minimize negative potential impacts on, e.g., allergy sufferers or agriculture.

Acknowledgements

Special thanks goes to T. Engel, A. Kunz, S. Brill, D. Kunkel, P. Hoor and T. Könemann for the helpful discussions and contributions to this work.

Conflict of interest

The authors declare no conflict of interest.

Author details

Daniel A. Pickersgill[1], Hartmut Müller[2*] and Viviane R. Després[3]

1 Max-Planck-Institute for Chemistry, Mainz, Germany

2 Mainz University of Applied Sciences, Mainz, Germany

3 Johannes-Gutenberg University, Mainz, Germany

*Address all correspondence to: hartmut.mueller@hs-mainz.de

References

[1] Pöschl U. Atmospheric aerosols: Composition, transformation, climate and health effects. Angewandte Chemie (International Ed. in English). 2005;**44**(46):7520-7540. DOI: 10.1002/anie.200501122

[2] Després Viviane R, Alex Huffman J, Burrows SM, Hoose C, Safatov AS, Buryak G, et al. Primary biological aerosol particles in the atmosphere. A review. Tellus B. 2012;**64**(0):296. DOI: 10.3402/tellusb.v64i0.15598

[3] Jones AM, Harrison RM. The effects of meteorological factors on atmospheric bioaerosol concentrations—A review. The Science of the Total Environment. 2004;**326**(1-3):151-180. DOI: 10.1016/j.scitotenv.2003.11.021

[4] Kellogg CA, Griffin DW. Aerobiology and the global transport of desert dust. Trends in Ecology & Evolution. 2006;**21**(11):638-644. DOI: 10.1016/j.tree.2006.07.004

[5] Morris CE, Sands DC, Bardin M, Jaenicke R, Vogel B, Leyronas C, et al. Microbiology and atmospheric processes. Research challenges concerning the impact of airborne micro-organisms on the atmosphere and climate. Biogeosciences. 2011;**8**(1):17-25. DOI: 10.5194/bg-8-17-2011

[6] Müller-Germann I, Vogel B, Vogel H, Pauling A, Fröhlich-Nowoisky J, Pöschl U, et al. Quantitative DNA analyses for airborne birch pollen. PLoS One. 2015;**10**(10):e0140949. DOI: 10.1371/journal.pone.0140949

[7] Fröhlich-Nowoisky J, Pickersgill DA, Despres VR, Pöschl U. High diversity of fungi in air particulate matter. Proceedings of the National Academy of Sciences of the United States of America. 2009;**106**(31):12814-12819. DOI: 10.1073/pnas.0811003106

[8] Draxler RR, Rolph GD. HYSPLIT (HYbrid Single-Particle Lagrangian Integrated Trajectory) NOAA Air Resources Laboratory, Silver Spring. 2003. Available from: http://www.arl.noaa.gov/ready/hysplit4.html, Zugriff 10/16

[9] Stein AF, Draxler RR, Rolph GD, Stunder BJB, Cohen MD, Ngan F. NOAA's HYSPLIT atmospheric transport and dispersion Modeling system. Bulletin of the American Meteorological Society. 2015;**96**(12):2059-2077. DOI: 10.1175/BAMS-D-14-00110.1

[10] Stohl A, Forster C, Frank A, Seibert P, Wotawa G. Technical note: The Lagrangian particle dispersion model FLEXPART version 6.2. Atmospheric Chemistry and Physics. 2005;**5**:2461-2474

[11] Rew R, Davis G, Emmerson S, Davies H, Hartnett E, Heimbigner D, et al. What is NetCDF? 2016. Available from: http://www.unidata.ucar.edu/software/netcdf/docs/index.html, Zugriff: 10/16

[12] Consortium for Small-scale Modelling. General Description of the COSMO-Model. 2011. Available from: http://www.cosmo-model.org/content/model/general/default.htm, Zugriff 10/16

[13] Brus DJ et al. Statistical mapping of tree species over Europe. European Journal of Forest Research. 2012;**131**(1):145-157

[14] CORINE Land Cover. Corine Land Cover 2006 raster data. 2006. Available from: http://www.eea.europa.eu/data-and-maps/data/corine-land-cover-2006-raster-3

[15] Environmental Systems Research Institute, Inc. What is ArcPy? 2016.

Available from: http://pro.arcgis.com/en/pro-app/arcpy/get-started/what-is-arcpy-.htm, Zugriff 10/16

[16] Paukstadt A. Extraktion meteorologischer Daten aus NetCDF für räumliche Analysen. Geoinformatik und Vermessung, Hochschule Mainz, master thesis; 2015. 110 pages. unpublished

[17] Stull R. An Introduction to Boundary Layer Meteorology. Alphen aan den Rijn, Netherlands: Kluwer Academic Publishers; 1988

[18] Seinfeld, Pandis. Atmospheric Chemistry and Physics. Hoboken, New Jersey, United States: Wiley Interscience; 2006

Chapter 5

The Color of Water from Space: A Case Study for Italian Lakes from Sentinel-2

Claudia Giardino, Kerttu-Liis Kõks, Rossano Bolpagni, Giulia Luciani, Gabriele Candiani, Moritz K. Lehmann, Hendrik Jan Van der Woerd and Mariano Bresciani

Abstract

Lakes are inestimable renewable natural resources that are under significant pressure by human activities. Monitoring lakes regularly is necessary to understand their dynamics and the drivers of these dynamics to support effective management. Remote sensing by satellite sensors offers a significant opportunity to increase the spatiotemporal coverage of environmental monitoring programs for inland waters. Lake color is a water quality attribute that can be remotely sensed and is independent of the sensor specifications and water type. In this study we used the Multispectral Imager (MSI) on two Sentinel-2 satellites to determine the color of water of 170 Italian lakes during two periods in 2017. Overall, most of the lakes appeared blue in spring and green-yellow in late summer, and in particular, we confirm a blue-water status of the largest lakes in the subalpine ecoregion. The color and its seasonality are consistent with characteristics determined by geomorphology and primary drivers of water quality. This suggests that information about the color of the lakes can contribute to synoptic assessments of the trophic status of lakes. Further ongoing research efforts are focused to extend the mapping over multiple years.

Keywords: chromaticity, multispectral sensors, optical remote sensing, inland waters, mapping, Sentinel-2, Italy, lakes

1. Introduction

Freshwater constitutes only 3% of the Earth's water resource, but only 1% is available as surface water in lakes and rivers, while the remainder is frozen in glaciers and ice caps or stored underground. Lakes represent a valuable source of water for consumption and irrigation and provide a variety of key services such as food provision, energy generation, transportation, recreation, and tourism. Lakes are essential components of the hydrological and biogeochemical cycles due to their basic ability to store, retain, clean, and provide water [1]. Lake waters also contribute to support the agricultural sector and livestock to feed the 7 billion of people on our planet [2].

Lake ecosystems are under pressure from various human impacts as well as climate change [3]. They are sensitive to a range of stressors operating at global, regional, and local scales [4] whose impacts manifest in eutrophication,

IntechOpen

proliferation of toxic algae, increase in turbidity, loss of aquatic benthos, and harmful effects on health for both animals and humans [5]. Significant effort is often devoted to monitor for changes, to the restoration of impacted systems, and to the preservation of healthy lakes. For example, in Europe, the need for having "[...] a coherent and comprehensive overview of water status within each river basin district" was defined by the Water Framework Directive (WFD) [6], setting out the requirements for the monitoring of the status of surface waters with the main objective of maintaining "good" and non-deteriorating status for all waters.

Earth observation (EO) techniques with optical sensors have been used for many decades to support timely and frequent acquisition of synoptic lake water quality information [7 and reference herein]. In recent years, EO has become an operational tool to support traditional measurements providing, at a relatively low cost and for some bio-geophysical parameters, information on surface water status to support a variety of applications [e.g., 8, 9]. EO systems measuring water quality typically are multispectral radiometers which might be grouped by their characteristic spatial and spectral resolution. Spatial resolution (the area on the ground covered by each pixel) is of particular importance for remote sensing of inland waters [10] as it determines the minimum size of lakes visible by each satellite. Four groupings of satellite sensors can currently be distinguished: ocean color (e.g., Sentinel-3 OLCI or MODIS, with pixels of about 300–1000 m), multispectral sensors (e.g., Landsat or Sentinel-2, with pixels of 10–30 m), imaging spectrometers (e.g., Hyperion or PRISMA, with a pixel size of 30 m, but coverage is not global unlike the previous missions), and geostationary platforms (e.g., GOCI, with a 500 m pixel size). Ocean color sensors provide better data for aquatic applications because they have more and narrower spectral bands and higher signal-to-noise ratios, but multispectral sensors are often the only choice for inland water applications because their finer spatial resolution can resolve smaller water bodies [10]. Multiple sensors might be used for improving the resolutions as in [11].

After processing of the light measured by a satellite sensor at the top of the atmosphere by removing light scattered by the atmosphere, stray light from adjacent pixels and specular reflection from the water surface physical and biochemical parameters of lakes can be estimated using several methods. Parameters that can be estimated include turbidity, photosynthetic biota (e.g., phytoplankton, macrophytes, and cyanobacteria), colored dissolved organic matter (CDOM, e.g., humic and fulvic substances), and suspended non-algal particulate matter (e.g., detritus from land). Lakes are complex ecosystems relative to oceanic waters due to the large variety and range of concentrations of living and nonliving material [12]. This complexity also applies to the optical properties, i.e., the spectral characteristics of absorption and scattering of light, of the constituents of lake water [13, 14], and, therefore, their estimation in lakes is extremely challenging. For example, if one component (e.g., CDOM) dominates the others (e.g., phytoplankton), it may mask the signature of the other components in the reflectance spectrum and reduce the accuracy of determining their concentrations. Due to this optical complexity, most algorithms for the retrieval of biogeochemical parameters are tailored to specific lakes and are not applicable to systems with optical properties different to those used for their development [e.g., 15, 16].

When research activities are focusing on mapping water quality in lakes from national to global scales, simpler yet robust approaches might be therefore strategically adopted. Among those, the methods estimating the color of water as perceived by the human eye show promise, because it does not rely on knowledge on inherent optical properties and concentrations of water components. Although perceived color is not unambiguously related to quantitative water quality attributes such as clarity, the phytoplankton, suspended matter, and CDOM, the color of water can be seen as a water quality attribute in its own right with the advantage of intuitive meaning in public perception.

The Commission Internationale de l'Éclairage (CIE) [17] mathematically defines color by weighting the reflectance spectrum of an object with three mixing curves, or chromaticity curves, each specifying the respective sensitivity of the human eye to one of the primary colors. To adapt this definition to the spectral bands of satellite sensors, several methods have been developed starting from the use of Forel-Ule (FU) scale, a historical standard recently recalibrated [18]. More recently, van der Woerd and Wernand [19] developed an algorithm to derive the hue angle consistently from different ocean color and multispectral sensors. Hue angle can be thought of as the pure color most closely resembling the true color of natural waters.

Several studies have used color analysis for a variety of applications in different aquatic ecosystems, including oceans and lakes. For example, [20] used chromaticity coordinates to prove the capability of Landsat-5 in assessing water quality changes from the pelagic to the coastal zone in Lake Garda (Italy). Wang et al [21] assessed the trophic state of global inland waters using a MODIS-derived Forel-Ule index finding that oligotrophic large lakes are concentrated in plateau regions in central Asia and South America, while eutrophic large lakes are concentrated in central Africa, eastern Asia, and mid-northern and southeast North America. In New Zealand, [22] calculate the color of water on almost 45,000 observations from 1486 lakes over 4 years. A preliminary exploratory analysis suggests that both geophysical and anthropogenic factors, such as catchment land use, provide environmental control of lake color and are promising avenues for future analysis. Lastly, [23] revealed that subtropical oceans will get bluer as fewer phytoplanktons are able to survive in its waters, while green regions at the poles will turn greener as warming waters become more habitable for them.

In this study, the method developed in [24] is adopted to calculate the color of Italian lakes based on multispectral Sentinel-2 images, whose 10-m spatial resolution allowed us to observe 170 lakes of the country. We follow [22] to analyze and classify lake colors from two different periods in 2017 for seasonal variations and patterns related to geomorphology and other primary drivers of water quality.

2. Matcrials and methods

2.1 Study area

About 2000 lakes are known in Italy, and ~500 of those have a surface area greater than 0.2 km^2 (400 of which are freshwater bodies and 100 brackish water bodies) [25]. The lakes are diverse systems with a plethora of values, including biodiversity, water provision, recreation, and landscape. For example, the volcanic-lake district located between Lazio and Basilicata administrative regions has 80% of the deep lakes within the Mediterranean coastal region holding 94% of the freshwater in central and southern Italy [26].

Lakes in Italy have different origins and features. Alpine lakes are generally small, fed by meltwater, and are normally located at altitudes above 2000 m a.s.l. where they occupy basins carved by glaciers. The deep subalpine lakes—the largest in Italy—occupy deep elongated valleys shaped by the erosive action of glaciers during the last glacial period. The debris left by ice on the edge of the plain forms the so-called morainic amphitheaters that, like the case of Lake Garda, still mark the southern limit of these water basins. The moraine lakes are entirely enclosed by hills formed by glacial deposits on the border between the Prealps and the Po Plain (e.g., lakes Viverone, Varese, Pusiano). The barrier lakes are formed following the obstruction of a river valley due to a landslide or the accumulation of alluvial sediments; examples are Lake Alleghe (landslide) and Lake Levico (accumulation of sediments). Volcanic lakes, mainly found in central Italy, feature an almost circular shape. Their formation is mainly related to subsidence and caldera formation during the final stages of volcanic activity [27]; examples are Lake Bolsena and Lake Bracciano. Alluvial lakes located in Central Apennines are formed by the filling of depressions originated by the raising of the Apennine chain (e.g., Lake Trasimeno). Other types of lakes include coastal and artificial ones.

Italy's overall lacustrine water volume is about 146 billion m^3, with seven large lakes (Garda, Maggiore, Como, Bolsena, Iseo, Bracciano, and Monte Cotugno) representing more than 97% of this amount. A major part of these lakes is located in the northern sector of the Italian Peninsula (along the Alpine range), although the Mediterranean regions are characterized by a high number of artificial lakes mainly supporting drinking or irrigation purposes. The morphology of the lakes is diverse with surface areas ranging between 3.4 and 370 km^2 (lakes Comabbio and Garda, respectively), maximum depths ranging between 2 and 410 m (Lesina and Como), and altitudes ranging between 0 and 507 m a.s.l. (lakes Lesina and Varano, and Vico).

Since 1997, a systematic investigation of morphological, physical, chemical, and biological features of the main lakes (with areas >0.2 km^2) has been implemented under the Project LIMNO. This project has the objective of developing a territorial information system for the interdisciplinary study of Italian lake environments. It consists of a database focused on morphometric, chemical, and biological data of water and sediments and the geographic information system tool (GIS LIMNO), which also includes thematic information on land use.

A major outcome of this project is the ability to analyze the physical and chemical variables for time trends in many lakes, especially the subalpine ones. For Lake Pusiano, it was revealed that the total phosphorus (TP) concentration, after having increased up to 200 μg/L (i.e., hypereutrophic) around the middle of the 1980s of the last century, has undergone a constant decline, down to the value of 58 μg/L in 2004. In other cases, however, opposite trends were observed. For example, Lake Garda exhibited TP concentrations in the range of 15 (1990s) to 34 μg/L during the 2004 circulation. In general, TP concentration shows higher values in lakes located at altitudes lower than 1000 m a.s.l. (median = 43 μg/L), while, for high-altitude lakes, this value never exceeds 4 μg/L. A similar trend has been also detected for total alkalinity (TAlk), with the highest values at low-altitude lakes (TAlk = 2.65 meq/L) and lowest values at high-altitude lakes (TAlk = 0.40 meq/L). This trend is also reflected by pH, which shows the minimum values at the highest altitudes. To sum up, the collected evidence has confirmed a considerable reduction in the maximum values of nutrients and contaminants even if data has often verified an increase in their basal levels.

2.2 Sentinel-2 data and processing

Sentinel-2 is a multispectral imaging mission of the Copernicus program. The mission that is funded by ESA Member States and the European Commission consists of twin satellites, the Sentinel-2A and Sentinel-2B, launched on 23 June 2015 on 7 March 2017, respectively.

Sentinel-2A and Sentinel-2B carry the Multispectral Imager (MSI), a push-broom sensor designed and built by Airbus Defense and Space, France. MSI has 13 spectral bands, ranging from the visible to the shortwave infrared (443–2190 nm) [28], with a swath width of 290 km and spatial resolutions of 10, 20, and 60 m (**Table 1**).

	S2A		S2B		
Band number	**Central wavelength (nm)**	**Bandwidth (nm)**	**Central wavelength (nm)**	**Bandwidth (nm)**	**Spatial resolution (m)**
b1	442.7	21	442.2	21	60
b2	492.4	66	492.1	66	10
b3	559.8	36	559.0	36	10
b4	664.6	31	664.9	31	10
b5	704.1	15	703.8	16	20
b6	740.5	15	739.1	15	20
b7	782.8	20	779.7	20	20
b8	832.8	106	832.9	106	10
b8a	864.7	21	864.0	22	20
b9	945.1	20	943.2	21	60
b10	1373.5	31	1376.9	30	60
b11	1613.7	91	1610.4	94	20
b12	2202.4	175	2185.7	185	20

Table 1.
Nominal settings of Sentinel-2A/Sentinel-2B MSI with band number, central wavelength, band width, and pixel size/resolution (source ESA).

By providing spatial resolution on the order of tens of meters and spectral bands comparable to the Operational Land Imager on Landsat 8 (and imagers on previous Landsat missions back to Landsat-5), Sentinel-2 is becoming to be considered as a key sensor for mapping lakes [10], which are often too small for ocean color sensors largely used in water quality studies [e.g., 21]. Then, considering the capacity of revisiting the same area every 5 days (2–3 days toward mid to high latitudes because of the overlap of the paths), Sentinel-2 is also useful for tracking changes over time scales of weeks. Therefore, in the last years, a number of lake studies have been developed with Sentinel-2 [e.g., 29–34].

In our study, 45 Sentinel-2A and Sentinel-2B MSI images were chosen, 22 during the spring (end of March to end of May) and the remaining acquired between late August and the end of September (late summer). Images were selected based on clear sky conditions and low glint contamination. Level-2C standard products were downloaded via the Copernicus Open Access Hub. The level-2C standard product is atmospherically corrected using the Sen2Cor [35]. Although the level-2C products rely on an atmospheric correction scheme not specifically designed for retrieving water leaving reflectance, it was recently demonstrated that its accuracy was better for inland than for coastal waters [36]; moreover level-2C MSI data have been used both in lake [37] and shallow water [38] applications. The MSI bands 1–5 were resampled at 10 m and then converted into remote sensing reflectance (*Rrs*) by dividing level-2C reflectance by π. The remaining spectral bands were not used as chromaticity which is entirely determined by light in the visible part of the spectrum. Finally, imagery data in *Rrs* units were imported into a GIS environment for clipping to vector outlines of the lakes listed in the geodatabase of the Italian Institute for Environmental Protection and Research (ISPRA). The 10-m spatial resolution of Sentinel-2 allowed us to consider 170 lakes down to a minimum size of 0.3 km^2. For each lake, the *Rrs* values were extracted from a square area avoiding islands and shallow waters, thus reducing the chance for mixed land-water pixels and bottom effects. The area used corresponds to a pixel window ranging from 3-by-3 to 90-by-90 and from smaller to larger lakes.

The chromaticity coordinates x, y, and z from MSI-derived *Rrs* data were computed by normalizing the individual tristimulus values X, Y, and Z:

$$x = \frac{X}{X+Y+Z};\ y = \frac{Y}{X+Y+Z};\ z = \frac{Z}{X+Y+Z};\ \text{with}\ x+y+z=1 \tag{1}$$

X, Y, and Z were computed as a linear weighted sum of MSI's five *Rrs* bands in the visible part of the spectrum (cf. **Table 1**) according to [24] (Eqs. (2)-(5)):

$$X = 8.356\,Rrs(b1) + 12.040\,Rrs(b2) + 53.696Rrs(b3) + 32.087Rrs(b4) + 0.487Rrs(b5) \tag{2}$$

$$Y = 0.993\,Rrs(b1) + 12.040\,Rrs(b2) + 53.696Rrs(b3) + 32.087Rrs(b4) + 0.177Rrs(b5) \tag{3}$$

$$Z = 43.487\,Rrs(b1) + 61.055\,Rrs(b2) + 1.778Rrs(b3) + 0.015Rrs(b4) \tag{4}$$

The x and y pairs were then plotted in the typical horseshoe-shaped chromaticity diagram (locus), where the center of the chromaticity diagram is the "white point" at which $x = y = z = 1/3$.

Any pair of x and y coordinates was then converted to hue angle (α). This is the angle between the line drawn from the white point to the x, y coordinate and the x-axis in anticlockwise direction. α was computed by using the four-quadrant arctangent function atan2 in MATLAB according to [22] (Eq. (5)):

$$\alpha = arctan\,(y - 1/3, x - 1/3) modulus 2\pi \tag{5}$$

The final step was the computation of dominant wavelength (λ_d). λ_d is the wavelength marked along the locus, and it is found as the intersection of the line drawn from the white point through the x, y coordinates.

3. Results and discussion

For each lake, x and y are plotted in the chromaticity diagram, commonly used to illustrate the color space and the range of colors in the sample. **Figure 1** depicts the natural color of our 170 lakes for each acquisition period in 2017. In spring, the colors of lakes are aligned elongated to a region spanning from blue toward green-orange; in late summer, the extent of the point cloud is greater and extends further into orange-red.

The optical properties of clean water are dominated by absorption and scattering by water molecules whose spectral dependence produces a blue reflectance spectrum. Therefore, the common perception that blue is "clean" is often true, while moving toward green, yellow, orange, and red, the optical effects of the other water components, such as phytoplankton, CDOM, and non-algal particle, become predominant. However, a simple back calculation from color to the direct causes of color change, e.g., proliferating phytoplankton or increasing sediment resuspension, is not possible. Nevertheless, any changes from blue can be reasonably attributed to decreasing water purity and is often also associated with a reduction in water clarity.

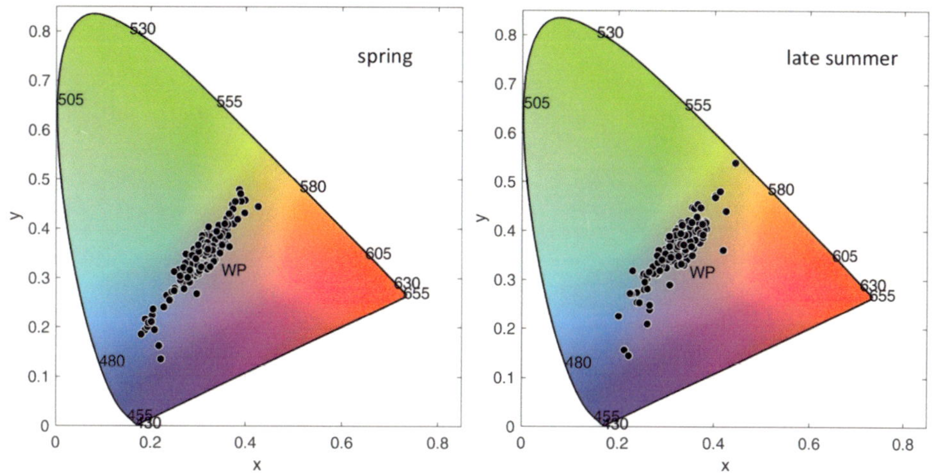

Figure 1.
Chromaticity diagram showing the color of water of 170 lake observations determined from Sentinel-2A/Sentinel-2B MSI data of 2017. Data related to spring are shown on the left, whereas those observed in late summer are shown on the right. The white point (WP: x = y = 1/3) is indicated as reference.

The frequency distribution of the dominant wavelength for the 170 lakes for both periods is plotted in **Figure 2**. Both histograms show a bimodal frequency distribution. In spring, most observations are in the blue-green part of the spectrum, with a secondary mode at green-yellow wavelengths. Vice versa, in late summer, most observations are in the green-yellow part of the spectrum and the secondary mode in the blue-green. To explain these changes, the lakes have been clustered according to three λ_d classes, defined as follows: blue (λ_d < 495 nm), green (495 nm < λ_d < 560 nm), and yellow (λ_d > 560 nm). In spring, 43% of lakes were classified as blue, 35% as green, and the last 22% as yellow. Moving toward late summer, most of the lakes were green (42%), then yellow (33%), and the remaining 25% as blue.

Of the 170 lakes, 96 did not show any transition from one color class to another, while 13 lakes moved from blue to yellow, showing a major change of optical properties also likely associated with a reduction in water clarity. The remaining 61 lakes showed smaller transitions to the neighboring color: 45 from blue to green or from green to yellow. The other 16 lakes showed transitions in the opposite direction, from green to blue or from yellow to green, suggesting improving water clarity from spring to late summer.

The geographic distribution of the three color classes is presented in **Figure 3**. Subalpine lakes in the northern part of the country including the largest lakes of the country (lakes Garda and Maggiore of 370 and 210 km^2, respectively) are distinctly blue in the spring. This lake district represents more than 80% of the total Italian lacustrine volume and is therefore of great interest. Moving from spring to late summer, a change of color toward green and yellow was observed in many of these lakes. Notably, the largest of these lakes, e.g., Lake Garda, remained blue.

A similar change is occurring in Sardinia, the second largest island of the country. Blue lakes turn yellow and green from spring to late summer. In contrast, only few lakes in Sicily show color transitions, and green and yellow colors prevail. Along the peninsula, more lakes are also blue during the spring than in late summer. However, a geographic gradient is seen in that summertime greening or yellowing is more common in the southern half of the peninsula.

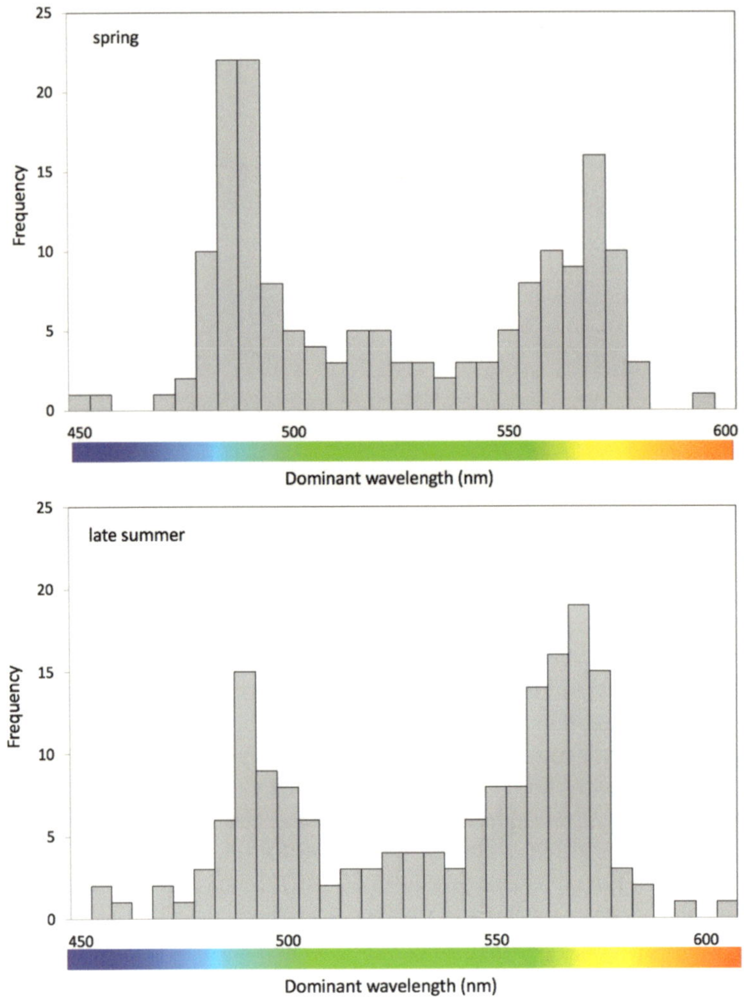

Figure 2.
Frequency histogram of dominant wavelength for the 170 lakes: on top, spring data; on bottom, late summer data.

To investigate these trends further, the lakes were split into four clusters according to latitude: northern (with latitudes >44°N), central (with latitudes in the range 44–41°N), southern lakes (with latitudes <44°N), and separately the lakes of Sardinia (**Figure 4**). Similar to general trends observed at the national scale, a progressive increase in λ_d was recorded moving from north to south in both the seasons. This is not surprising as the wide latitudinal range of Italy (~38–47°N) encompasses marked climatic, geological, topographic, and land use gradients.

A possible explanation, regardless of physical and chemical differences between lakes, is that the Mediterranean lakes are characterized by an advance of the growing season compared to northern ones. This may translate into an early start of the algal growth with significant effects on the color of the water. Consequently, it is quite natural to guess higher levels of productivity (colors basically more green-yellow) for southern lakes, as described by [39] at a global scale. Additionally, the differences between the two main Mediterranean and Italian islands, Sicily and Sardinia, are probably due to the geological and climatic differences between the two islands [40] and are likely exacerbated by the fact that their lakes are largely artificial reservoirs with site-specific trophic drivers.

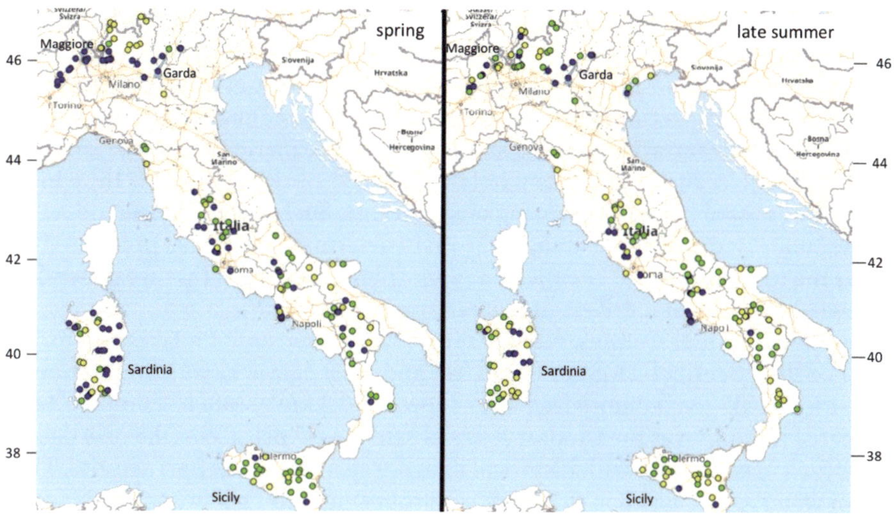

Figure 3.
Geographic distribution of lakes, colored according to their dominant wavelength: on left, spring observations; on right, summer observations. The latitudes in degrees are indicated as reference.

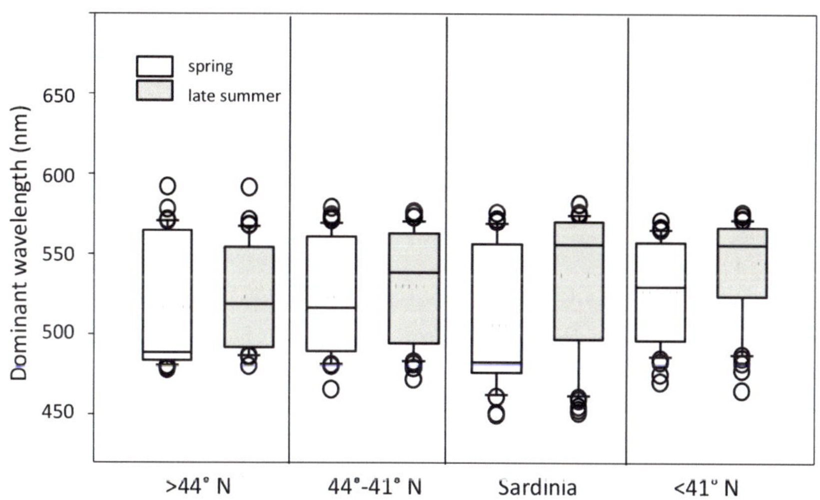

Figure 4.
Box plot depicting dominant wavelengths for spring (white) and autumn (gray) periods for northern (>44°N), central (44–41°N), Sardinian, and southern (<41°N) lakes.

4. Conclusion

In this study, the color of water, a simple and straightforward water quality attribute quantitatively described in terms of dominant wavelength, was retrieved from Sentinel-2A and Sentinel-2B MSI data. The method allowed us to map the color of 170 Italian lakes in two periods during 2017.

The results revealed a general increase in λ_d moving from north to south (in the range ~38–47°N) and from spring to late summer. This could be put in relation to the macroclimatic differences associated with the latitudinal gradient under investigation. Moreover, the observed trends suggest that the investigation of drivers of water chromaticity can contribute to fundamental understanding of lake water quality. This represents an opportunity for water managers who have to act under the dramatic effects of climate change on water availability and quality.

Our work shows that color observations are an efficient means to capture an intuitive water quality attribute at spatial and temporal scales practically impossible to achieve using ground-based observations. Further investigations are required to relate color of water to trophic status and traditional water quality metrics such as chlorophyll *a* concentration and suspended particulate matter. Such relationships most likely require the classification of lakes into bio-optical types [14, 16] which can also be assisted by remote sensing observations. Such knowledge would help to better understand and disentangle the main determinants of lake productivity such as the role of physical, chemical, and morphometric traits that are generally acknowledged as pivotal drivers of primary production [26].

For more than four decades, satellite sensors have been used for lake monitoring, and since 2015, Sentinel-2 MSI provides free and open data at a spatial resolution suitable for small- to medium-sized lakes (down to 0.3 km^2). MSI has similar spectral and spatial resolution as the Landsat series of satellites which allows the new data to be analyzed in continuity with historical imagery spanning back four decades. The color of water as calculated in this work is a promising water quality attribute for time series analysis as it does not rely on algorithms depending on inherent optical properties that have to be calibrated with field observations. While the present study only looked at two seasons in the same year, a long-term analysis could investigate the timing of summertime greening of the lakes in response to climatic forcing mechanism.

Ongoing research is focused on extending the color mapping over past observations. Future applications of chromaticity analysis are promising as each Sentinel-2 satellite has a 7-year lifetime design, and they are planned to be replaced in the framework of ESA's Copernicus Program in 2022–2023 by new identical missions. This ensures continuity of the data record to 2030 and provides the opportunity for lake water quality monitoring for decades from now.

Acknowledgements

Sentinel-2 images were downloaded from Copernicus Open Access Hub. Morphometric characteristic of the Italian lakes were obtained CNR-LIMNO Project (http://www.ise.cnr.it/limno/limno.htm). We also thank Pietro Alessandro Brivio who has shared, with some of us, his knowledge on lake chromaticity at the end of the 1990s.

Conflict of interest

No potential conflict of interest is reported by the authors.

Author details

Claudia Giardino[1*], Kerttu-Liis Kõks[2], Rossano Bolpagni[1,3], Giulia Luciani[1], Gabriele Candiani[1], Moritz K. Lehmann[4], Hendrik Jan Van der Woerd[5] and Mariano Bresciani[1]

1 Institute for Electromagnetic Sensing of the Environment, National Research Council of Italy (CNR-IREA), Milan, Italy

2 ERASMUS + at Institute for Electromagnetic Sensing of the Environment, National Research Council of Italy (CNR-IREA), Milan, Italy

3 Department of Chemistry, Life Sciences and Environmental Sustainability, Parma University, Parma, Italy

4 Xerra Earth Observation Institute, Alexandra, New Zealand

5 Institute for Environmental Studies (IVM), Water and Climate Risk, VU University Amsterdam, Amsterdam, The Netherlands

*Address all correspondence to: giardino.c@irea.cnr.it

References

[1] Wetzel RG. Limnology: Lake and River Ecosystems. 3rd ed. San Diego, CA, USA: Academic Press; 2001. 1006 p

[2] Pimentel D, Berger B, Filiberto D, Newton M, Wolfe B, Karabinakis E, et al. Water resources: Agricultural and environmental issues. Bioscience. 2004;**54**(10):909-918. DOI: 10.1641/0006-3568(2004)054[0909:WRAAEI]2.0.CO;2

[3] Hartmann DL, Tank AMGK, Rusticucci M. IPCC Fifth Assessment Report, Climate Change 2013: The Physical Science Basis. IPCC Ar5, 5, 2013. pp. 31-3

[4] Niinemets Ü, Kahru A, Nõges P, Tuvikene A, Vasemägi A, Mander Ü, et al. Environmental feedbacks in temperate aquatic ecosystems under global change: Why do we need to consider chemical stressors? Regional Environmental Change. 2017;**17**(7): 2079-2096. DOI: 10.1007/s10113-017-1197-2

[5] Paerl HW, Paul VJ. Climate change: Links to global expansion of harmful cyanobacteria. Water Research. 2012;**46**:1349-1363. DOI: 10.1016/j.watres.2011.08.002

[6] The European Parliament and the Council of the European Union. Directive 2000/60/EC of the European Parliament and of the council of 23 October 2000 establishing a framework for community action in the field of water policy. Official Journal of the European Communities. 2000;**L327**:1-72

[7] Palmer SC, Kutser T, Hunter PD. Remote sensing of inland waters: Challenges, progress and future directions. Remote Sensing of Environment. 2015;**157**:1-8. DOI: 10.1016/j.rse.2014.09.021

[8] Tyler AN, Hunter PD, Spyrakos E, Groom S, Constantinescu AM, Kitchen J. Developments in earth observation for the assessment and monitoring of inland, transitional, coastal and shelf-sea waters. The Science of the Total Environment. 2016;**572**:1307-1321. DOI: 10.1016/j.scitotenv.2016.01.020

[9] Dörnhöfer K, Oppelt N. Remote sensing for lake research and monitoring—Recent advances. Ecological Indicators. 2016;**64**:105-122. DOI: 10.1016/j.ecolind.2015.12.009

[10] Olmanson LG, Brezonik PL, Bauer ME. Remote sensing for regional lake water quality assessment: Capabilities and limitations of current and upcoming satellite systems. In: Advances in Watershed Science and Assessment. Cham: Springer; 2015. pp. 111-140. DOI: 10.1007/978-3-319-14212-8_5

[11] Hestir EL, Brando VE, Bresciani M, Giardino C, Matta E, Villa P, et al. Measuring freshwater aquatic ecosystems: The need for a hyperspectral global mapping satellite mission. Remote Sensing of Environment. 2015;**167**:181-195. DOI: 10.1016/j.rse.2015.05.023

[12] Lee Z, Carder KL, Hawes SK, Steward RG, Peacock TG, Davis CO. Model for the interpretation of hyperspectral remote-sensing reflectance. Applied Optics. 1994;**33**(24):5721-5732. DOI: 10.1364/AO.33.005721

[13] Mouw CB, Greb S, Aurin D, DiGiacomo PM, Lee Z, Twardowski M, et al. Aquatic color radiometry remote sensing of coastal and inland waters: Challenges and recommendations for future satellite missions. Remote Sensing of Environment. 2015;**160**:15-30. DOI: 10.1016/j.rse.2015.02.001

[14] Spyrakos E, O'Donnell R, Hunter PD, Miller C, Scott M, Simis SG, et al. Optical types of inland and coastal

waters. Limnology and Oceanography. 2018;**63**(2):846-870. DOI: 10.1002/lno.10674

[15] Odermatt D, Gitelson A, Brando VE, Schaepman M. Review of constituent retrieval in optically deep and complex waters from satellite imagery. Remote Sensing of Environment. 2012, 2012;**118**:116-126. DOI: 10.1016/j.rse.2011.11.013

[16] Eleveld M, Ruescas A, Hommersom A, Moore T, Peters S, Brockmann C. An optical classification tool for global lake waters. Remote Sensing. 2017;**9**(5):420. DOI: 10.3390/rs9050420

[17] CIE. Commission Internationale de l'Éclairage Proceedings. Cambridge, UK: Cambridge University Press; 1932

[18] Novoa S, Wernand MR, Van der Woerd HJ. The Forel-Ule scale revisited spectrally: Preparation protocols, transmission measurements and chromaticity. Journal of the European Optical Society—Rapid Publications. 2013;**8**:1-8. DOI: 10.2971/jeos.2013.13057

[19] Van der Woerd H, Wernand M. True colour classification of natural waters with medium-spectral resolution satellites: SeaWiFS, MODIS, MERIS and OLCI. Sensors. 2015;**15**(10):25663-25680. DOI: 10.3390/s151025663

[20] Zilioli E, Brivio PA, Gomarasca MA. A correlation between optical properties from satellite data and some indicators of eutrophication in Lake Garda (Italy). Science of the Total Environment. 1994;**158**:127-133. DOI: 10.1016/0048-9697(94)90052-3

[21] Wang S, Li J, Zhang B, Spyrakos E, Tyler AN, Shen Q, et al. Trophic state assessment of global inland waters using a MODIS-derived Forel-Ule index. Remote Sensing of Environment. 2018;**217**:444-460. DOI: 10.1016/j.rse.2018.08.026

[22] Lehmann M, Nguyen U, Allan M, van der Woerd H. Colour classification of 1486 lakes across a wide range of optical water types. Remote Sensing. 2018;**10**(8):1273. DOI: 10.3390/rs10081273

[23] Duncombe J. The deep blue sea is getting bluer. Eos. 2019;**100**. DOI: 10.1029/2019EO115971. Available from: https://eos.org/articles/the-deep-blue-sea-is-getting-bluer

[24] van der Woerd H, Wernand M. Hue-angle product for low to medium spatial resolution optical satellite sensors. Remote Sensing. 2018;**10**(2):180. DOI: 10.3390/rs10020180

[25] Pagnotta R, Passino R. Research programmes in limnological field in Italy and effects of the current environmental legislation. Memorie dell'Istituto Italiano di Idrobiologia. 1992;**50**:19-28

[26] Azzella MM, Rosati L, Iberite M, Bolpagni R, Blasi C. Changes in aquatic plants in the Italian volcanic-lake system detected current data and historical records. Aquatic Botany. 2014;**112**:41-47. DOI: 10.1016/j.aquabot.2013.07.005

[27] Mosello R, Arisci S, Bruni P. Lake Bolsena (Central Italy): An updating study on its water chemistry. Journal of Limnology. 2004;**63**:1-12. DOI: 10.4081/jlimnol-20004.1.

[28] Drusch M, Del Bello U, Carlier S, Colin O, Fernandez V, Gascon F, et al. Sentinel-2: ESA's optical high-resolution mission for GMES operational services. Remote Sensing of Environment. 2012;**120**:25-36. DOI: 10.1029/2005RG000183.

[29] Dörnhöfer K, Göritz A, Gege P, Pflug B, Oppelt N. Water constituents and water depth retrieval from Sentinel-2A—A first evaluation in an oligotrophic lake. Remote Sensing.

2016;**8**(11):941. DOI: 10.3390/rs8110941

[30] Martins V, Barbosa C, de Carvalho L, Jorge D, Lobo F, Novo E. Assessment of atmospheric correction methods for Sentinel-2 MSI images applied to Amazon floodplain lakes. Remote Sensing. 2017;**9**(4):322. DOI: 10.3390/rs9040322

[31] Bresciani M, Cazzaniga I, Austoni M, Sforzi T, Buzzi F, Morabito G, et al. Mapping phytoplankton blooms in deep subalpine lakes from Sentinel-2A and Landsat-8. Hydrobiologia. 2018;**824**(1):197-214. DOI: 10.1007/s10750-017-3462-2

[32] Pinardi M, Bresciani M, Villa P, Cazzaniga I, Laini A, Tóth V, et al. Spatial and temporal dynamics of primary producers in shallow lakes as seen from space: Intra-annual observations from Sentinel-2A. Limnologica. 2018;**72**:32-43. DOI: 101016/j.limno.2018.08.002

[33] Ansper A, Alikas K. Retrieval of Chlorophyll a from Sentinel-2 MSI data for the European Union water framework directive reporting purposes. Remote Sensing. 2019;**11**(1):64. DOI: 10.3390/rs11010064

[34] Fritz C, Kuhwald K, Schneider T, Geist J, Oppelt N. Sentinel-2 for mapping the spatio-temporal development of submerged aquatic vegetation at Lake Starnberg (Germany). Journal of Limnology. 2019;**78**(1):71-91. DOI: 10.4081/jlimnol.2019.1824

[35] Main-Knorn M, Pflug B, Louis J, Debaecker V, Müller-Wilm U, Gascon F. Sen2Cor for Sentinel-2. In: Image and Signal Processing for Remote Sensing XXIII. International Society for Optics and Photonics; October 2017. 2017. 10427:1042704. DOI: 10.1117/12.2278218

[36] Warren MA, Simis SGH, Martinez-Vicente V, Poser K, Bresciani M, Alikas K, et al. Assessment of atmospheric correction algorithms for the Sentinel-2A MultiSpectral imager over coastal and inland waters. Remote Sensing of Environment. 2019;**225**:267-289. DOI: 10.1016/j.rse.2019.03.018

[37] Toming K, Kutser T, Laas A, Sepp M, Paavel B, Nõges T. First experiences in mapping lake water quality parameters with Sentinel-2 MSI imagery. Remote Sensing. 2016;**8**(8):640. DOI: 10.3390/rs8080640

[38] Evagorou E, Mettas C, Agapiou A, Themistocleous K, Hadjimitsis D. Bathymetric maps from multi-temporal analysis of Sentinel-2 data: The case study of Limassol, Cyprus. Advances in Geosciences. 2019;**45**:397-407. DOI: 10.5194/adgeo-45-397-2019

[39] Adrian R, O'Reilly CM, Zagarese H, Baines SB, Hessen DO, Keller W, et al. Lakes as sentinels of climate change. Limnology and Oceanography. 2009;**54**(6part2):2283-2297. DOI: 10.4319/lo.2009.54.6_part_2.2283

[40] Dinelli E, Lima A, Albanese S, Birke M, Cicchella D, Giaccio L, et al. Major and trace elements in tap water from Italy. Journal of Geochemical Exploration. 2012;**112**:54-75. DOI: 10.1016/j.gexplo.2011.07.009

Chapter 6

Bacubirito: An Outstanding Cosmic Sample on Earth

Emiliano Terán

Abstract

The Bacubirito meteorite, although largely forgotten by the scientific community after its excavation in 1902, remains an incredible artifact and inspired generations of Mexican scientists. It is the fifth largest meteorite in the world and the longest with a length of 4.16 m. Using the Monte Carlo method, an innovative geometrical model and scanner, we have been the first to calculate the precise dimensions and mass (21 tons) of Bacubirito meteorite's complex structure. Moreover, we are advocating that it be added to the list of world heritage sites due to its scientific, cultural, and historical significance in Mexico and the world.

Keywords: Bacubirito, iron meteorites, anomalous meteorite, tridimensional model, volume, mass

1. Introduction

Meteorites have provided tremendous amounts of information about the planetary system and the cores of planets or star while fascinating societies for centuries. These pieces of asteroids or fragmented planets are composed of rock, iron, or a mixture of both. Iron meteorites were part of the nucleus of planetoids or asteroids, rock from the surface and the rock-iron from the intermediate zone. Mexico's large surface area contains a considerable number of valuable meteorites, and in the last 100 years, scientific research has evolved to study them. For instance, the Mexican Allende meteorite was used to calculate the age of our solar system [1].

The Bacubirito meteorite is a famous Mexican meteorite found in a small town named Camichín in the mountain range of Sinaloa in Northwestern Mexico, see **Figure 1**. The name comes from the closest town Bacubirito, and it has been the source of scientific research in Mexico for the last century. This enormous meteorite has an outstanding length of 4.16 m and a mass close to 20 tons.

We hope that the research presented here will also support our case for the meteorite's inclusion in the list of Heritage Sites in the United Nations Educational, Scientific, and Cultural Organization (UNESCO) and demonstrate its value to Mexican scientific research and the world's knowledge of meteorites. From our point of view, it meets the following criteria [2]:

> *"VII Containing extreme natural phenomena or areas of exceptional natural beauty or esthetic importance; VIII To be one of the representative examples of important historical stages of the history of the earth including testimonies of life, geological processes creating geological formations, or significant physiographic or geomorphic characteristics."*

Figure 1.
Map of México with the state of Sinaloa in detail. Culiacán the capital city of Sinaloa and the town of Bacubirito are shown as well. The meteorite is currently located in Culiacán.

Further arguments for its significance will appear in this chapter, and we will also discuss its impact on Mexican science in addition to the innovative method utilized to determine its mass.

2. Mexican scientific interest in meteorites

The magnitude of the surface of Mexico (1,964,375 km^2) allowed the discovery of a great number of meteorites. We can calculate the probability of meteorite impacts [3] on the Mexican territory with the cumulative distribution $N(d)$ of diameters of Earth-Crossing Asteroids (ECAs) derived by Poveda et al. in order to estimate the frequency of collisions of meteoroids with cars and aircraft. A meteorite with a 9-cm diameter strikes a car with an expected frequency of once every 5 years according to the available data for meteorite strikes. Mexico's large surface area also increases the frequency of meteorite strikes, although the correlation of meteorite impact and geographical location remains unclear. Even so, iron meteorites can have catastrophic effects on Earth. However, they were once the main source of metal for tools and weapons in ancient times [3], and in spite of the great amount of metallic metals in Mexico, there is no record of these tools in museums, perhaps taken during the Spanish conquest [2].

The early Mexican Scientific Community advocated the study and the catalog of meteorites and other geological features, and they successfully petitioned the government to protect and categorize them as a national heritage. For this reason, any of these objects appear in museums or public institutions. The field of geology and the mining industry promoted scientific development in Mexico in the nineteenth century. Scientists trained in these disciplines began to dedicate themselves to the study of the earth and established the procedures to protect, study, and catalog the meteorites.

Two mining engineers [4], Antonio Del Castillo (1820–1895) and Mariano Santiago de Jesús de la Bárcena y Ramos (Mariano Barcenas for future reference) (1842–1899) featured prominently in the initial study of meteorites in Mexico and guaranteed legal protection for these objects while advancing research in the field. Castillo, an outstanding academic, created schools and institutes about geology and geography (*Escuela Practica de Minas* and *Consejo de Minería y obras Publicas and Colegio de Mineria*). He also wrote the *Catalogue descriptif des météorites du Mexique de 1889* (a 200-page work discussing meteorites in Mexico), and he published his excavations of Chihuahua's meteorites (displayed in the entrance of the National School of Engineers in 1893). Although Castillo describes the Bacubirito meteorite in this work, he did excavate or visit it, and thus, he relied on an estimation to calculate its mass. Moreover, he was a member of the federal congress, and in a letter sent on May 7, 1889, he proposed legislation to protect all Mexican celestial objects, including meteorites, from private ownership and prohibited their destruction, exportation, alienation, and required their conservation. Thus, Bacubirito is the property of the Republic of Mexico and cannot be exported or sold, the current constitution of Mexico still contains this legislation. Moreover, he influenced the interest of other scientists to study meteorites, particularly in his former students.

Mariano Barcenas was one of the most internationally well-known Mexican scientists and one of the most prominent students of Del Castillo. He combined an outstanding academic career with a successful public life as an entrepreneur and politician. In the congress of the Academy of Natural Sciences in Philadelphia in 1876 [5], he made the Bacubirito meteorite internationally known. Barcenas studied this meteorite by means of a sample that the governor of the state, Eustaquio Buelna-Perez from 1871 to 1875. Afterward, he obtained the composition of the meteorite and gave a mass estimation, though not precisely, since no other information was available without excavation. Despite the interest around meteorite, it remained buried for many more years, mainly due to the difficulty in arriving to the Camichin.

Professor Henry Augustus Ward (1834–1906), from the Rochester Academy of Science of New York and meteorite collector, excavated the meteorite in 1902 in Mexico [6, 7]. He describes in detail the hard work to arrive to Bacubirito and how the 20 individuals were required to unearth this enormous celestial piece, see **Figure 2**. He also reported the object's dimensions: 13 feet, 1-inch length (3.96 m), 6 feet, 2 inches wide (1.8 m), and 5 feet, 4 inches in depth (1.52 m) and mass of 50 tons. He estimated the object's mass assuming a cubic shape (although the meteorite is highly irregularly shaped). Despite of the affection that Camichin's population had for the meteorite, it was moved to Constitution Civic Center in Culiacan (see **Figure 3**), the capital of the state of Sinaloa in 1959, a 245.5-km route, in order to facilitate the object's study.

Figure 2.
Unearthing of the specimen in 1902 by local individual.

Figure 3.
Photograph of the former location of the meteorite in the constitution Civic Park (Parque Civico Constitución) in 1959.

Figure 4.
Photograph of the meteorite in Culiacan in 2018 [8].

Professor Vagn Fabritius Buchwald (1929–current) from the Department of Mechanical Engineering in the Technical University of Denmark visited the meteorite in 1978. He evaluated the physical properties of Bacubirito meteorite and determined its volume and mass. He described the width, which varied between 100 and 185 cm and a depth between 30 and 75 cm. The intricate shape and ear-like curve made it difficult to calculate the exact volume. However, he used a photographical method to arrive at a volume of 2.80 m^3 ± 10%, corresponding to a weight of 22 tons: a good approximation with the tools available at the time. In 1993, the meteorite was relocated to its current position in Science Center of Sinaloa in Culiacan (**Figure 4**).

A new generation of scientist in Mexico is analyzing and promoting Bacubirito meteorite [8, 9]. They have obtained new measurements that place it as the longest in the world (4.1 m). They have also promoted it in the State Congress of Sinaloa in 2018 and declared that the meteorite should be considered a cultural and historical patrimony of the state. This was achieved with the support of Dr. Victor Antonio Corrales-Burgueño (1954–today), a congressman and former Head Dean of the Autonomous University of Sinaloa (UAS). Thus, with this legislation, we hope to have the meteorite declared to Human Heritage, so that it will be included in this esteemed list and take place as an "exceptional and universal value."

One of the strongest arguments to promote this initiative is the studies that we have achieve to measure the weight of this metal giant. As we have seen in its description, its size has made this difficult to establish with accuracy. Notwithstanding, advances in techniques and methods have enabled us to establish the final weight of this specimen. In the following section, we explain this in detail.

3. Geometrical model of Bacubirito to assess its mass

As stated formerly, the meteorite has suffered classification changes in size over time and is considered anomalous in density, which may be due to a lack of detailed investigations. Back in 1975, Buchwald indicated that only a few studies existed on the meteorite, and little has been done 40 years later. There have only been rough estimates of properties, such as its mass, dimensions, and densities, as noted from the large mass variations reported. Moreover, for most of the estimations, uncertainties have not been provided. Accurate measures of geometrical parameters, the mass, and statistics of the regmaglypts were determined from the elaboration of a three-dimensional model of Bacubirito.

The importance of studying this meteorite can be understood in terms of the role it plays on the entry dynamics of the object. If we consider the theoretical prediction for speed given by Regan and Anandakrishnan [10], two important variables appear namely the drag coefficient C_d and cross-sectional area A that are determined by its geometry. Another important dynamic variable that can be investigated is the mass using the ablation model in Revelle [11], and from the two parameters just mentioned, an estimate of the object mass can be obtained before entry. Finally, given that the meteorite shape is nonspherical and assuming that it

Figure 5.
Perspectives of the Bacubirito meteorite.

was not greatly modified during entrance, we can use our geometrical model and study the drag force for an asymmetrical object as in Leith [12]. However, these studies are beyond the scope of the present investigation, and for this research, we will limit ourselves with the model acquisition and parameters previously cited.

The Bacubirito meteorite has a rather complex shape (see **Figure 5**) and large weight (**Tables 1** and **2**), and thus, determining the surface shape of Bacubirito can be used to infer important information about its passage through the atmosphere and learn more about the meteorite itself. The geometric model can also provide reliable classification for meteorites, and comparisons between this and other recovered pieces contain additional details [9]. This model allowed us to calculate some of its important features: geometric parameters, mass, and statistics of regmaglypts. First, we generated the model, and we defined and calculated the geometric features. An analysis of reported densities to obtain a precise mass estimate was performed. Then, we reported the size and depth of the regmaglypts, and finally, we obtained a precise geometrical model of the meteorite.

Under these restrictions, we have chosen to use a portable scanner to determine the model. This instrument allows for a vastly accurate estimation of the relative positions of surface points on a rigid object and is highly suited from a practical perspective—the outdoor location of the meteorite and its resting on a high position are benefits in this case. This device has been used in civil engineering applications, where detailed, precise and three-dimensional representations are required. Being a highly characterized instrument, it allows us to reduce the uncertainty of our results.

The scanner used was the Leica Nova MS50. This equipment radiates and collects a laser beam, which can either directly interact with the object to measure it or by means of a prism. The direct operation mode is mostly employed since it offers a higher resolution and precision to acquire the 3-d points. According to an extensive characterization from the manufacturer (see Leica Nova MS50 Datasheet), it has an angular accuracy of =1″ in horizontal and vertical angular measurements, a

Meteorite	Country	Dimensions [m × m × m]	Mass [tons]
Hoba	Africa	2.7 × 2.7 × 0.9	60
Campo del Cielo (El Chaco)	Argentina	–	31
Cape York (Ahnighito)	Greenland	3.25 × 2.1 × 1.6	30.88
Cape York (Agpalilik)	Greenland	2.1 × 2.0 × 1.5	20.14
Bacubirito	México	4.1 × 1.8 × 0.2*	19.43±0.51*

*Revised table from Buchwald [13]. *Our measurement.*

Table 1.
List of the largest meteorites of the world.

Author	Year	Mass [tons]
Ward	1902	50
Angerman	1903	25
Merrill	1929	20
La Paz	1973	27
Buchwald	1975	22 ± 10%
Sanchez-Rubio	2001	19

Table 2.
Reported Bacubirito meteorite mass estimates until year 2001 (see Refs.).

linear precision of l = 2–3 mm + 2 ppm, and collects up to 1000 pts./s. The scanner can measure objects up 1 km away. Our measurements, however, were conducted at distances of about 15 m (see **Figure 2**). We placed the scanner in five different positions around the meteorite in order to reach every spot on its surface. For the inaccessible regions, a prism was employed. We finished with a model of 1,812,875 points on the meteorite's surface.

Before proceeding to computing the geometric parameters, we propose a Monte Carlo simulation to obtain even more precise positions and evaluate the uncertainty of our results from the single series of measurements. A classical reference to the Monte Carlo method and its applications in many areas is found in Rubinstein and Kroese [14]. With this goal in mind, we first develop a simple probabilistic model and later explain its practical implementation. We start assuming that the uncertainties in our measured positions are random, namely, that our calibrated instrument presents negligible systematic deviations as supported by the small parts per million on its linear accuracy. Furthermore, both distance and angle measurements present systematic-free errors that follow a Gaussian distribution. Now, the propagation error in position due to the angular uncertainty is negligible (≪1 mm) for our short separation distances and angular precision, and we conclude that this will follow a Gaussian distribution. From this discussion, the deviation $\overrightarrow{\delta r_i}$ of a measured position with respect to the instrument $\vec{r_i}$ can be written as $\overrightarrow{\delta r_i} = \overrightarrow{\lambda r_i}$, with $\vec{r_i}$ a unit vector (see **Figure 6**), where the λ parameter k is the random variable in the Monte Carlo simulation with a Gaussian distribution centered at zero and σ = 6 mm/2. This information is valid for both the characterization provided by the manufacturer (see datasheet) and the measurement technique in which the instrument was positioned in different locations with uncertainties lower than 4 mm.

The application of the method consists of taking the ith position measurement $\vec{r_i}$ and then, a particular value for λ is generated based on the distribution described above. We then add $\overrightarrow{\delta r_i}$ and repeat the process for each of the measurements. These amounts are obtained from one experiment from which the desired parameters can be extracted. The experiment is repeated many times for a simulation and is averaged over the intermediate results leading to a convergence.

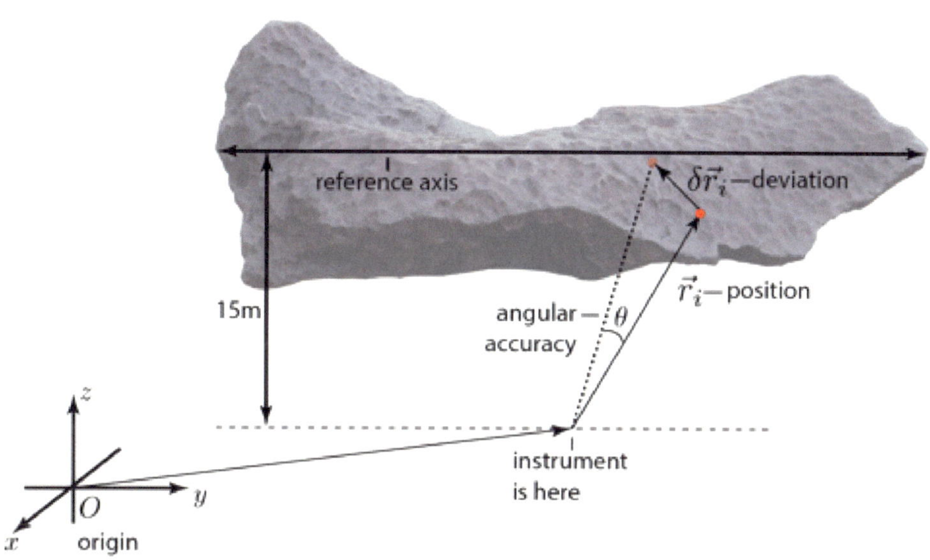

Figure 6.
Angular and linear deviations of the surface points due to intrinsic measurements uncertainty.

Finally, we start determining Bacubirito's dimensions. We define its length as the separation between the two farthest-apart points and fix an axis onto these points. We next define the width as the distance between the two farthest-away points that are both perpendicular to the axis and lie on a plane containing the axis. Similarly, the thickness is defined as the largest distance between the two points that define a line orthogonal to that plane and that pass through the axis. A sketch is shown in **Figure 7**.

In the next step of experiment, we estimate the meteorite volume by combining the Monte Carlo simulation with Gauss' Theorem. A Monte Carlo experiment is first employed. Gauss' Theorem is then used to estimate the volume enclosed by the simulated surface defined by a Delaunay triangulation. Thus, we obtained the associated volumes for each simulation run.

We continued with the analysis of the meteorite's densities reported so far and its mass determination. The percentages of its chemical elements are shown in **Table 3** (see Ref. [9]). Those values were measured in different samples from Bacubirito. On average, the measurements should be an approximation of the full meteorite's concentrations and densities. We obtained the average density of 7.7250 ± 0.2061 g/cm^3 from the mentioned table and with the aid of the volume calculated the mass.

The diameter and depth of the regmaglypts were the next determined quantities using the model. We defined a regmaglypt's diameter as that of the best-fit circle for the corresponding pit mouths. The depth is the difference between the bottom and the top. In order to characterize the distribution of regmaglypts, three zones around the meteorite were selected for presenting contrasting structures as shown in **Table 1**. The diversity in structures was assessed statistically using the Kolmogorov-Smirnov test for a thorough and rigorous treatment of the technique. To this end,

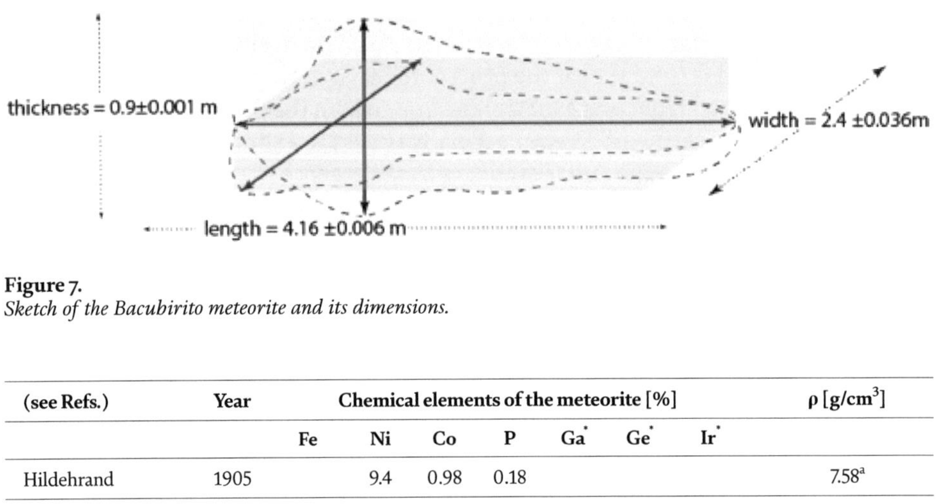

Figure 7.
Sketch of the Bacubirito meteorite and its dimensions.

(see Refs.)	Year	Chemical elements of the meteorite [%]							ρ [g/cm^3]
		Fe	Ni	Co	P	Ga*	Ge*	Ir*	
Hildehrand	1905		9.4	0.98	0.18				7.58[a]
Moore & Lewis	1968		9.78	0.76	0.16	17.7	31.9	4.9	7.97
Scott	1973		9.62			17.7	31.9	4.9	7.62
Wasson, pers. comm.	1968		9.62			17.7	31.9	4.9	7.62
La Paz	1973	88.94	6.98	0.21	0.15				7.64
Average ± std. dev.									7.7250 ± 0.2061

[a]*Quantity determined by Cohen according to La Paz (La Paz 1973).*
**Gallium, Germanium and Iridium are presented in a ppm scale.*

Table 3.
Reports of elements concentrations for the Bacubirito meteorite.

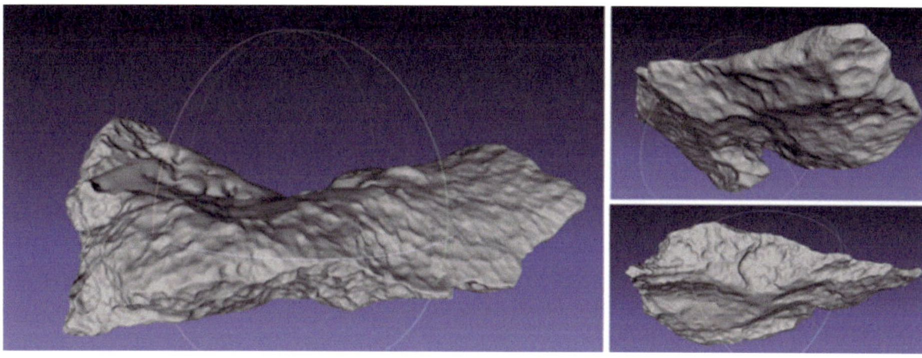

Figure 8.
Renders of the Bacubirito meteorite model.

we retrieved diameters and depths for each region from the model and took each as a random variable. The null hypothesis that any two of the regions originate from a same distribution for either variable was then evaluated.

3.1 Results and discussion

Renders of our tridimensional model are given in **Figure 8** and can be observed in a high level of detail. A precise retrieval of any geometrical parameters and its derivatives is also possible. We focus in this section on the dimensions, mass, and statistics of the regmaglypts.

The basic defining parameters of Bacubirito are the dimensions and their uncertainties depicted in the last column of **Table 4**. Contrasting differences are observed with respect to one of the first studies performed by Ward and the latest one by Buchwald. While estimates of geometric parameters of Bacubirito have been reported previously, different studies exhibited noticeable changes. We made the formal definition for consistency among studies and determine them precisely from the model. Therefore, we can confirm the old claim that Bacubirito is the longest meteorite in the world, with a total length of 4.130 ± 0.005 m (see **Table 4**).

The previous estimate of volume was based on photographs by Buchwald [13] resulting in 2.8 ± 10% m^3. This was very good approximation considering the prevalent instrumentation and data analysis techniques of the time. The volume calculated from a number of simulations through the Monte Carlo method is:

$$V = 2.5151 \pm 0.0005 \ m^3 \tag{1}$$

This represents a novel rigorous calculation of the volume of the meteorite. The remaining two parameters give the reader the idea of the aspect ratio of the body and its importance among meteorites in the world.

Dimensions	Ward [m]	Buchwald [m]	This work [m]
Width	1.88	1.00–1.85	2.053 ± 0.005
Thickness	1.63	0.30–0.75	0.911 ± 0.005
Length	3.99	4.25	4.130 ± 0.005

Table 4.
Geometrical parameters of the Bacubirito meteorite.

We emphasize that the density of points considered here (a total of 1,812,875) reflects an upper bound for an instrument with uncertainties in the millimeter range. An increase in the number of points would result in separation distances along the meteorite surface in the range of μm, which is well beyond the maximum precision of the device.

The mass is calculated from averaging the densities in **Table 3** and our volume estimation.

$$m = 19.429 \pm 0.51 \text{ tons} \tag{2}$$

We obtained a mass uncertainty of around 3%, and this was consistent with the estimations made by Buchwald and Sanchez-Rubio et al. [15]. Furthermore, we can note that these densities represent the main uncertainty, and hence, we recommend improving their estimation.

Finally, an important feature of iron meteorites is their regmaglypts (**Figure 9**). Those meteorite traits provide important information related to its fall. For example, according to Lin and Qun [16], larger pits correspond to zones where the surface vector is more aligned with the meteorite velocity vector. The statistics of the regmaglypts are depicted in **Table 5**, which demonstrates different areas on the meteorite that exhibit contrasting average diameters and depths. To confirm those differences, a Kolmogorov-Smirnov test shows a p value of <0.01 among the regions. Consequently, each region has a characteristic structure and implies different origins.

These data obtained from Bacubirito distinguish it from other meteorites in the world in length and size and for previous generations of scholars presented a significant challenge.

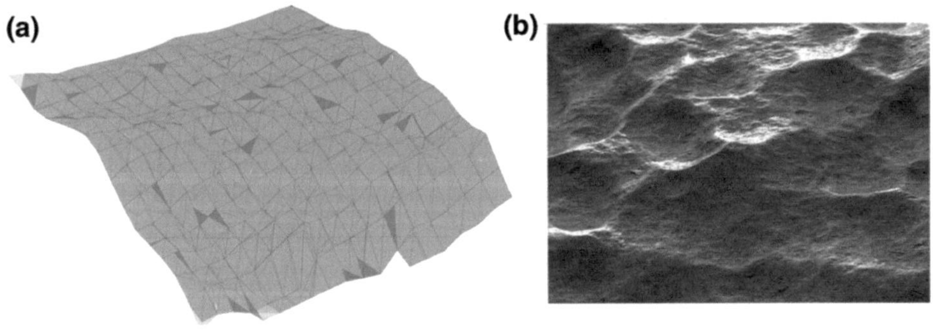

Figure 9.
(a) Render and (b) photography of Bacubirito.

Region	Diameter [cm]	Depth [cm]
Front	8.0 ± 2.3	1.0 ± 0.3
Rear	12.6 ± 2.6	1.4 ± 0.3
Below	10.9 ± 2.6	2.2 ± 0.4

Table 5.
Diameter and depth of the meteorite's regmaglypts.

4. Conclusions

In this work, we obtained a geometrical model that determined the mass of the Bacubirito meteorite and its other geometrical features. Our results are listed as follows:

1. The mass has a value of 19.43 ± 0.51 tons. The results are both the geometric dimensions and mass. These were achieved through state-of-the-art analysis techniques and equipment, considerable improvement on previous estimations.

2. The volume of the meteorite is 2.5151 ± 0.0005 m^3. We utilized the Monte Carlo method to simulate a series of measurements for a tridimensional rigid object in order to reduce time and expenses related to the repetition of in-field measurements.

3. The maximum length of the Bacubirito is 4.130 ± 0.005 m—the longest known in the world, and we also obtained estimations of the meteorite's main dimensions.

4. The depth and width of the regmaglypts of three regions were retrieved. The regions were found to exhibit different structures. The zone labeled as below is most likely to have been exposed directly to the atmosphere in the ablation altitudes.

The study also supports our proposal to include the Bacubirito meteorite on the list of the World Heritage Sites for its tremendous impact on culture, science, history, and geological studies. In addition to the impressive physical characteristics, mass, and length of the meteorite, the early Mexican scientific community developed through its research of the object and developed a series of legislative initiatives to conserve celestial objects. Moreover, local residents of Bacubirito and Sinaloa believe that the object represents national heritage and is a source of local pride.

Author details

Emiliano Terán[1,2]

1 Physics Department, Autonomous University of Sinaloa, Culiacán, Sinaloa, Mexico

2 Astronomical Observatory, Autonomous University of Sinaloa, Cosalá, Sinaloa, Mexico

*Address all correspondence to: eteran@uas.edu.mx

References

[1] Amelin Y, Krot A. Pb isotopic age of the Allende chondrules. Meteoritics & Planetary Science. 2007;**42**(7-8):1321-1335

[2] Leon-Portilla M. The Aztec Account of the Conquest of Mexico. Beacon Press; 2011

[3] Jambon A. Bronze age iron: Meteoritic or not? A chemical strategy. Journal of Archaeological Science. 2017;**88**:47-53

[4] Morelos-Rodriguez L. La geología mexicana en el siglo XIX. Plaza y Valdés; 2012. 356 p

[5] Mariano B. On certain Mexican meteorites. Proceedings of the Academy of Natural Sciences of Philadelphia. 1876;**28**:122-126

[6] Ward HA. On Bacubirito, the great meteorite of Sinaloa, Mexico. Science. 1902;**16**(401):395-398

[7] Morelos-Rodriguez L. 125 años de las Meteoritas en el Palacio de Minería. Facultad de Ingeniería; 2018

[8] Terán-Bobadilla E, Abundis-Patiño JH, Añorve-Solano C, Moraila-Valenzuela CR, OrtegaGutierrez F. Bacubirito: The longest meteorite in the world. Astronomy & Geophysics. 2018;**59**(2):2.30-22.1

[9] Terán-Bobadilla E, Abundis-Patiño JH, Añorve C, Moraila CR, Ortega-Gutiérrez F, Aragón-Calvo MA. On a novel geometric analysis of the Bacubirito meteorite. Earth, Moon, and Planets. 2017;**120**(2):101-111

[10] Regan FJ, Anandakrishnan SM. Dynamics of Atmospheric Re-entry. American Institute of Aeronautics and Astronautics; 1993

[11] Revelle DO. A quasi-simple ablation model for large meteorite entry: Theory vs observations. Journal of Atmospheric and Terrestrial Physics. 1979;**41**(5):453-473

[12] Leith D. Drag on nonspherical objects. Aerosol Science and Technology. 1987;**6**(2):153-161

[13] Buchwald VF. Handbook of Iron Meteorites. Berkeley: University of California Press; 1975

[14] Dirk K, Rubinstein R. Simulation and the Monte Carlo Method. John Wiley & Sons; 2016

[15] Sánchez-Rubio GR-KR, Lozano-Santa Cruz R, Medina-Malagón S, Flores-Gutiérrez D. In: UNAM, editor. Las meteoritas de México. Instituto de Geología; 2001

[16] Lin TC, Qun P. On the formation of regmaglypts on meteorites. Fluid Dynamics Research. 1987;**1**(3):191-199

Chapter 7

Spatial Coordinate Transformations with Noisy Data

Christopher Kotsakis

Abstract

The parametric transformation of spatial coordinates between different reference frames is traditionally dealt with a stepwise approach which imposes a suboptimal treatment in the presence of noisy data. The chapter explains briefly the drawbacks of this approach and then presents an alternative scheme for spatial coordinate transformations that improves the classic stepwise solution when using noisy coordinates of known stochastic structure. The proposed methodology is simple in principle, although its numerical implementation with nonlinear parametric models is a bit more involved and it relies on the joint least squares adjustment of the observed coordinates using their full stochastic model over all points of interest. The mathematical framework and the related properties of this "stacking" approach are presented in detail, along with a numerical example that demonstrates its feasibility for practical problems in geospatial applications.

Keywords: spatial adjustment, coordinate transformation, reference frames, nonlinear least squares, stacked Gauss-Helmert model, noise filtering

1. Introduction

Spatial coordinate-based positions often need to be transformed from their own reference frame to another reference frame by an analytic parametric model. This is a standard problem in several fields of geosciences and engineering, including geodesy [11, 18, 19, 30, 33], land surveying and cadastral planning [2, 8, 16, 32], cartography and digital mapping [1, 5, 28, 38], photogrammetry and remote sensing [22, 25, 39], robotics and computer vision [7, 13, 31], among others.

Spatial coordinate transformations (SCTs) are utilized in practice either explicitly for determining the unknown coordinates of scattered points in a desired frame from their observed coordinates in a different frame or implicitly in the context of more composite procedures such as the self-calibration of terrestrial laser scanners [20, 23, 26], the conflation of digital maps and geographical databases [2, 4, 6, 27], the reconstruction of 3D models from multi-sensor data [9, 29, 35], and the integration of aerial or satellite images in ground-based systems of geographic coordinates [14, 15, 21]. Various technical terms have actually been used with regard to SCTs in practical problems, for instance, *spatial adjustment*, *image registration*, *absolute orientation*, *geo-referencing*, and *frame transformation*, to name a few. Despite their linkage to different application fields, all these terms refer, more or less, to the same archetypical problem, that is, the optimal fusion of partially overlapping configurations of spatial points using their coordinate-based representations in separate frames and an application-specific model to describe their systematic differences. Although this general viewpoint includes also cases with raster-type and/or vector-type data in multiple frames, the present chapter is confined to the vectorized version of SCT problems between two different frames.

2. Problem statement and research motivation

The transformation of spatial coordinates between different reference frames using a parametric model is traditionally resolved via a two-step procedure. Firstly, the parameters of the adopted model are estimated by least squares adjustment (LSA) techniques on the basis of control points with known coordinates in both frames of interest. The estimated parameters are then used to transform any set of known coordinates from the original frame (also called source or initial frame) to the desired target frame. The second step is applied not only to control points but also to additional points whose spatial coordinates are originally known only in the initial frame. Their transformed coordinates are often the primary objective in many practical applications, whereas the post-fit residuals at control points are commonly used as quality metrics of the transformation process.

Despite its rational character, the stepwise methodology imposes a suboptimal treatment in the presence of noisy data. The reason is that the accuracy of the original coordinates in the initial frame is ignored during their transformation to the target frame. Indeed, the second step entails only the forward implementation of the parametric model without attempting to minimize or, at least, reduce the propagated random errors of the original coordinates. As shown in [18], the stepwise procedure may actually enhance the data noise into the final results, in the sense that the transformed coordinates in the target frame could become less accurate than the original coordinates in the initial frame.

An additional weakness of the traditional stepwise procedure is that the control and non-control points (abbreviated hereafter as CPs and NPs) are handled independently throughout the transformation process. Yet, the initial coordinates of these points are usually acquired from the same pre-analysis phase or observational procedure (e.g., geodetic network adjustment, map digitization, etc.), and therefore they are affected by common error sources. Due to their separate treatment, the cross-correlated part of the initial coordinate errors at CPs and NPs is always ignored, a fact that may weaken the accuracy of the transformed coordinates in the target frame.

The focus of the present research is the formulation of a single-stage estimation scheme that can improve the classic stepwise solution in SCT problems. The proposed scheme is based on a properly weighted least squares adjustment of all observed coordinates, using their known variances and covariances (CV) in the entire group of transformation points. This stacking approach permits the rigorous treatment of intra-frame error correlations among CPs and NPs, and it generally leads to higher-accuracy results for the transformed coordinates. Our contribution provides easy-to-use optimal estimators for the transformed coordinates under any parametric model, regardless of the structure of the error covariance matrices of the input data. A numerical example is also given at the end of the chapter to demonstrate the feasibility of the proposed methodology for practical applications.

3. The stepwise approach in spatial coordinate transformations

3.1 Preamble

The mathematical setting is based on the general case of nonlinear transformation models, in accordance to the vectorized expression:

$$X = f(X', \theta) \quad (1)$$

where X' and X contain the Cartesian coordinates for the same group of physical points (or homologous groups of points) in different reference frames. The spatial dimensionality does not need to be specified here, and it can refer to any case that occurs in practice. The vector θ represents the parameters of the transformation model which enable the coordinate mapping from the initial frame to the target frame.

In the following, Eq. (1) is considered as an exact formula for noise-free coordinates and provides the general framework for the LSA of observed coordinates in the involved frames. Simpler types of transformation models with joint or marginal linearity in X' and θ (e.g., errors-in-variables models, differential or close-to-identity models) can be also analyzed under the previous setting.

For the purpose of this contribution, the user's data shall consist of:

a. the observed coordinates for CPs and NPs in the initial frame (denoted by X' and Z', respectively);

b. the observed coordinates for CPs in the target frame (denoted by X); and

c. the error CV matrices of the previous vectors (denoted by $\Sigma_{X'}$, $\Sigma_{Z'}$, Σ_X).

An additional matrix of special importance is the cross-CV matrix $\Sigma_{X'Z'}$ which reflects the *intra-frame* error correlation between CPs and NPs, and it is totally ignored in the traditional stepwise procedure.

3.2 Estimation of transformation parameters

The first step refers to the estimation of the transformation parameters using a sufficient number of known CPs. Following a statistical estimation perspective, the optimal parameter values are obtained by solving the nonlinear LSA problem

$$\min_{\theta} \left(v_X^T \Sigma_X^{-1} v_X + v_{X'}^T \Sigma_{X'}^{-1} v_{X'} \right) \quad (2)$$

subject to

$$X + v_X = f(X' + v_{X'}, \theta) \quad (3)$$

where the vectors v_X and $v_{X'}$ represent the zero-mean random errors in the observed coordinates. After appropriate linearization, the above problem can be reduced to a linear LSA problem for the so-called Gauss-Helmert (GH) model [12, 17], and it leads to an iterative solution via successive refinements of the preliminary estimate:

$$\hat{\theta} = \theta_o + \left(J_\theta^T W J_\theta \right)^{-1} J_\theta^T W (X - f(X', \theta_o)) \quad (4)$$

$$W = \left(\Sigma_X + J_{X'} \Sigma_{X'} J_{X'}^T \right)^{-1} \quad (5)$$

where θ_o contains approximate values for the transformation parameters. The recursive updating of the previous solution is performed by the Newton-Gauss iteration algorithm in accordance to a more complex expression that will be presented later in this chapter. The matrices $J_{X'}$ and J_θ are the Jacobians with respect to the initial frame coordinates and the transformation parameters, that is,

$$J_{X'} = \frac{\partial f(X', \theta)}{\partial X'}, \quad J_\theta = \frac{\partial f(X', \theta)}{\partial \theta} \tag{6}$$

and they need to be re-evaluated at each iteration step using the adjusted values from the previous step. For more details on nonlinear least squares adjustment and iterative computational algorithms, the reader should consult the excellent treatise in [24] (see also [3, 10, 34]).

3.3 Determination of transformed coordinates

After estimating the transformation parameters, an additional step is required to complete the solution of the problem at hand, that is, the computation of the transformed coordinates in the target frame. This is performed by a simple forward evaluation of the transformation model at the CPs and NPs, using the respective nonlinear formulae:

$$\hat{X} = f\left(X', \hat{\theta}\right) \tag{7}$$

$$\hat{Z} = f\left(Z', \hat{\theta}\right). \tag{8}$$

Note that $\hat{\theta}$ corresponds to the estimated parameters from the first step, whereas X' and Z' refer to the observed coordinates in the initial frame. The following Jacobian matrices are also defined here (to be used later on):

$$J_{Z'} = \frac{\partial f(Z', \theta)}{\partial Z'}, \quad \tilde{J}_\theta = \frac{\partial f(Z', \theta)}{\partial \theta} \tag{9}$$

which differ from their previous counterparts in Eq. (6) as they refer to a separate group of points (NPs).

3.4 Deficiency of the stepwise approach

A number of drawbacks exist in the stepwise approach for SCT problems with noisy data. More specifically, (a) the noise of the original coordinates remains unfiltered during their transformation to the target frame, (b) the correlated errors in the original coordinates between CPs and NPs are not taken into account, and (c) the accuracy of the transformed coordinates is not optimized under any statistical principle. All these drawbacks relate to the same modeling deficiency that is summarized as follows: the observed coordinates in the initial frame are contaminated by random errors which remain uncontrolled during the second step of the transformation process, and they are fully absorbed by the transformed coordinates of CPs and NPs.

The aforesaid deficiency is irrelevant for practical applications only in two cases:

- if the sole objective is to determine a set of transformation parameters between different frames, without the need to perform any coordinate transformation at specific points; or

- if spatial objects (e.g., point cloud, network, digitized map) need to be transferred from an initial frame to another frame, without any "quality improvement" of their transformed coordinates.

However, if the user's goal is the optimal referencing of spatial objects with respect to a target frame, then the unfiltered data noise becomes a critical error source for SCT problems. This does not mean that the stepwise approach leads to wrong results, but it signifies that the composite estimators in Eqs. (7) and (8) do not provide an optimal solution of maximum accuracy for the transformed coordinates.

It is worth noting that the stepwise approach is not compelled to reproduce the prior reference coordinates of CPs in the target frame, that is, $\hat{\boldsymbol{X}} \neq \boldsymbol{X}$, even if these coordinates are perfectly known without any errors!

3.5 Best-fitting transformation solutions

In some cases, the estimation of transformation parameters is performed via the alternative nonlinear least squares principle [36, 37]:

$$\min_{\boldsymbol{\theta}} \| \boldsymbol{X} - \boldsymbol{f}(\boldsymbol{X}', \boldsymbol{\theta}) \|^2 \tag{10}$$

where $\|\cdot\|$ denotes the standard form of the Euclidean vector norm. The rationale of the above principle is to bring in the best alignment two different coordinate sets over a group of CPs, and it does not lead to the same parameter estimates as the statistical least squares formulation of Section 3.2. Their formal equivalency occurs if the known coordinates in the initial frame are treated as noiseless quantities and the respective coordinates in the target frame are affected by uncorrelated random errors of equal variance. Nevertheless, Eq. (10) has a strong geometrical significance, and it is often used in practice regardless of the noise characteristics of the available data.

If the transformation parameters are obtained by the alternative principle of Eq. (10), then it obviously holds that

$$\| \boldsymbol{X} - \hat{\boldsymbol{X}} \|^2 \rightarrow \min \tag{11}$$

which implies that the transformed coordinates of CPs will be optimally fitted, in a least squares sense, to their prior known values in the target frame. This best-fitting property does not enforce statistical optimality to the accuracy of the transformed coordinates—the latter will still absorb the entire observation noise according to Eqs. (7) and (8). Therefore, the point to be stressed here is that a high-quality transformation solution should not just rely on the fitting performance at CPs, but it has to exploit in an optimal sense the stochastic error model of the observed coordinates over all points of interest.

4. The stacking approach in spatial coordinate transformations

4.1 Theoretical aspects

A unified optimal solution for SCT problems can be obtained in a single stage through the rigorous combination of all available data. This requires the joint LSA of the nonlinear transformation equations:

$$\boldsymbol{X} = \boldsymbol{f}(\boldsymbol{X}', \boldsymbol{\theta}) \tag{12}$$

$$\boldsymbol{Z} = \boldsymbol{f}(\boldsymbol{Z}', \boldsymbol{\theta}) \tag{13}$$

which should be performed in a linearized context via the Newton-Gauss iteration method [3, 24, 34]. The algebraic setup of this stacking adjustment and the basic properties of the resulting estimators for the transformed coordinates are presented in this section.

4.1.1 Linearization

At first, we need to approximate the nonlinear Eqs. (12) and (13) by the truncated multivariate Taylor's series expansions:

$$\boldsymbol{X} = \boldsymbol{f}(\boldsymbol{X}'_o, \boldsymbol{\theta}_o) + \boldsymbol{J}_{X'}(\boldsymbol{X}' - \boldsymbol{X}'_o) + \boldsymbol{J}_{\theta}(\boldsymbol{\theta} - \boldsymbol{\theta}_o) \tag{14}$$

$$\boldsymbol{Z} = \boldsymbol{f}(\boldsymbol{Z}'_o, \boldsymbol{\theta}_o) + \boldsymbol{J}_{Z'}(\boldsymbol{Z}' - \boldsymbol{Z}'_o) + \tilde{\boldsymbol{J}}_{\theta}(\boldsymbol{\theta} - \boldsymbol{\theta}_o) \tag{15}$$

where $\boldsymbol{\theta}_o$ is a vector of approximate values for the transformation parameters and $\boldsymbol{X}'_o$, $\boldsymbol{Z}'_o$ are vectors of approximate coordinates for the respective points in the initial frame. Taking into account that the observables correspond to the coordinate vectors $\boldsymbol{X}$, $\boldsymbol{X}'$, and $\boldsymbol{Z}'$, the previous formulae should be further augmented as follows:

$$\boldsymbol{X} + \boldsymbol{v}_X = \boldsymbol{f}(\boldsymbol{X}'_o, \boldsymbol{\theta}_o) + \boldsymbol{J}_{X'}(\boldsymbol{X}' + \boldsymbol{v}_{X'} - \boldsymbol{X}'_o) + \boldsymbol{J}_{\theta}(\boldsymbol{\theta} - \boldsymbol{\theta}_o) \tag{16}$$

$$\boldsymbol{Z} = \boldsymbol{f}(\boldsymbol{Z}'_o, \boldsymbol{\theta}_o) + \boldsymbol{J}_{Z'}(\boldsymbol{Z}' + \boldsymbol{v}_{Z'} - \boldsymbol{Z}'_o) + \tilde{\boldsymbol{J}}_{\theta}(\boldsymbol{\theta} - \boldsymbol{\theta}_o) \tag{17}$$

where the added vectors $\boldsymbol{v}_X$, $\boldsymbol{v}_{X'}$, $\boldsymbol{v}_{Z'}$ denote the zero-mean random errors of the observed coordinates. The linearized expressions (16) and (17) can be equivalently written in the block-matrix form:

$$\begin{bmatrix} -\boldsymbol{J}_{\theta} & \boldsymbol{0} \\ -\tilde{\boldsymbol{J}}_{\theta} & \boldsymbol{I} \end{bmatrix} \begin{bmatrix} \boldsymbol{\theta} - \boldsymbol{\theta}_o \\ \boldsymbol{Z} \end{bmatrix} + \begin{bmatrix} \boldsymbol{I} & -\boldsymbol{J}_{X'} & \boldsymbol{0} \\ \boldsymbol{0} & \boldsymbol{0} & -\boldsymbol{J}_{Z'} \end{bmatrix} \begin{bmatrix} \boldsymbol{v}_X \\ \boldsymbol{v}_{X'} \\ \boldsymbol{v}_{Z'} \end{bmatrix} + \begin{bmatrix} \boldsymbol{X} - \boldsymbol{f}(\boldsymbol{X}'_o, \boldsymbol{\theta}_o) - \boldsymbol{J}_{X'}(\boldsymbol{X}' - \boldsymbol{X}'_o) \\ -\boldsymbol{f}(\boldsymbol{Z}'_o, \boldsymbol{\theta}_o) - \boldsymbol{J}_{Z'}(\boldsymbol{Z}' - \boldsymbol{Z}'_o) \end{bmatrix} = \begin{bmatrix} \boldsymbol{0} \\ \boldsymbol{0} \end{bmatrix} \tag{18}$$

which conforms to the usual structure of Gauss-Helmert linear models of statistical estimation theory [12, 17, 24]. Our objective here is to invert the above stacked system of the general form $\boldsymbol{Ax} + \boldsymbol{Bv} + \boldsymbol{w} = 0$ using the general least squares principle $\boldsymbol{v}^T \boldsymbol{P} \boldsymbol{v} = \min$, in conjunction with the data weight matrix:

$$\boldsymbol{P} = \boldsymbol{\Sigma}^{-1} = \begin{bmatrix} \boldsymbol{\Sigma}_X & \boldsymbol{0} & \boldsymbol{0} \\ \boldsymbol{0} & \boldsymbol{\Sigma}_{X'} & \boldsymbol{\Sigma}_{X'Z'} \\ \boldsymbol{0} & \boldsymbol{\Sigma}_{Z'X'} & \boldsymbol{\Sigma}_{Z'} \end{bmatrix}^{-1} \tag{19}$$

which reflects the total statistical accuracy of the observables. Note that the *inter-frame* correlations of the observed coordinates are assumed to be zero, whereas the *intra-frame* correlations between CPs and NPs are taken into account by the cross-CV matrix $\boldsymbol{\Sigma}_{Z'X'} (= \boldsymbol{\Sigma}^T_{X'Z'})$.

If applied under a proper iterative setting, the LSA of Eq. (18) leads to the sought optimal solution of the problem at hand. Specifically, the transformation parameters and the coordinates of NPs in the target frame are both contained into the "parameter vector" of the stacked GH-type model, and they can be directly obtained via the respective least squares estimator (see next section). On the other hand, the estimated coordinates of CPs in the target frame shall be deduced in an implicit way by correcting the observed values $\boldsymbol{X}$ for the effect of their random errors ($\boldsymbol{v}_X$) which are also estimable from the iterative least squares inversion of Eq. (18).

4.1.2 Optimal least squares estimators

By applying the general LSA solution of linear GH models (see [12, 17]) to the stacked system of Eq. (18) and after some extra lengthy derivations using analytic inversions of 2 × 2 block matrices, we obtain the explicit estimators for the transformation parameters:

$$\hat{\boldsymbol{\theta}} = \boldsymbol{\theta}_o + \left(\boldsymbol{J}_\theta^T \boldsymbol{W} \boldsymbol{J}_\theta\right)^{-1} \boldsymbol{J}_\theta^T \boldsymbol{W} \left(\boldsymbol{X} - \boldsymbol{f}(\boldsymbol{X}'_o, \boldsymbol{\theta}_o) - \boldsymbol{J}_{X'}(\boldsymbol{X}' - \boldsymbol{X}'_o)\right) \tag{20}$$

and for the coordinates of NPs in the target frame

$$\begin{aligned} \hat{\boldsymbol{Z}} = {} & \boldsymbol{f}(\boldsymbol{Z}'_o, \boldsymbol{\theta}_o) + \boldsymbol{J}_{Z'}(\boldsymbol{Z}' - \boldsymbol{Z}'_o) + \tilde{\boldsymbol{J}}_\theta\left(\hat{\boldsymbol{\theta}} - \boldsymbol{\theta}_o\right) \\ & + \boldsymbol{J}_{Z'} \boldsymbol{\Sigma}_{Z'X'} \boldsymbol{J}_{X'}^T \boldsymbol{W}\left(\boldsymbol{X} - \boldsymbol{f}(\boldsymbol{X}'_o, \boldsymbol{\theta}_o) - \boldsymbol{J}_{X'}(\boldsymbol{X}' - \boldsymbol{X}'_o) - \boldsymbol{J}_\theta\left(\hat{\boldsymbol{\theta}} - \boldsymbol{\theta}_o\right)\right) \end{aligned} \tag{21}$$

whereas the estimated errors for each subset of observed coordinates are given by the equation

$$\begin{bmatrix} \hat{\boldsymbol{v}}_X \\ \hat{\boldsymbol{v}}_{X'} \\ \hat{\boldsymbol{v}}_{Z'} \end{bmatrix} = \begin{bmatrix} -\boldsymbol{\Sigma}_X \\ \boldsymbol{\Sigma}_{X'} \boldsymbol{J}_{X'}^T \\ \boldsymbol{\Sigma}_{Z'X'} \boldsymbol{J}_{X'}^T \end{bmatrix} \boldsymbol{W}\left(\boldsymbol{X} - \boldsymbol{f}(\boldsymbol{X}'_o, \boldsymbol{\theta}_o) - \boldsymbol{J}_{X'}(\boldsymbol{X}' - \boldsymbol{X}'_o) - \boldsymbol{J}_\theta\left(\hat{\boldsymbol{\theta}} - \boldsymbol{\theta}_o\right)\right) \tag{22}$$

The auxiliary matrix $\boldsymbol{W}$ that appears in the previous equations was defined earlier in Section 3.2. Finally, if we combine the first error component from Eq. (22) with the basic formula $\hat{\boldsymbol{X}} = \boldsymbol{X} + \hat{\boldsymbol{v}}_X$, we get the estimated coordinates of CPs in the target frame:

$$\begin{aligned} \hat{\boldsymbol{X}} = {} & \boldsymbol{f}(\boldsymbol{X}'_o, \boldsymbol{\theta}_o) + \boldsymbol{J}_{X'}(\boldsymbol{X}' - \boldsymbol{X}'_o) + \boldsymbol{J}_\theta\left(\hat{\boldsymbol{\theta}} - \boldsymbol{\theta}_o\right) \\ & + \boldsymbol{J}_{X'} \boldsymbol{\Sigma}_{X'} \boldsymbol{J}_{X'}^T \boldsymbol{W}\left(\boldsymbol{X} - \boldsymbol{f}(\boldsymbol{X}'_o, \boldsymbol{\theta}_o) - \boldsymbol{J}_{X'}(\boldsymbol{X}' - \boldsymbol{X}'_o) - \boldsymbol{J}_\theta\left(\hat{\boldsymbol{\theta}} - \boldsymbol{\theta}_o\right)\right) \end{aligned} \tag{23}$$

To facilitate a comprehensive analysis of the stacking approach, it is useful to rewrite Eqs. (21) and (23) in the *combined Kalman-like form*:

$$\begin{bmatrix} \hat{\boldsymbol{X}} \\ \hat{\boldsymbol{Z}} \end{bmatrix} = \begin{bmatrix} \frown{\boldsymbol{X}} \\ \frown{\boldsymbol{Z}} \end{bmatrix} + \begin{bmatrix} \boldsymbol{J}_{X'} \boldsymbol{\Sigma}_{X'} \boldsymbol{J}_{X'}^T \\ \boldsymbol{J}_{Z'} \boldsymbol{\Sigma}_{Z'X'} \boldsymbol{J}_{X'}^T \end{bmatrix} \underbrace{\left(\boldsymbol{\Sigma}_X + \boldsymbol{J}_{X'} \boldsymbol{\Sigma}_{X'} \boldsymbol{J}_{X'}^T\right)^{-1}}_{\boldsymbol{W}} \left(\boldsymbol{X} - \frown{\boldsymbol{X}}\right) \tag{24}$$

where the auxiliary terms $\frown{\boldsymbol{X}}$ and $\frown{\boldsymbol{Z}}$ are strictly given by the expressions

$$\begin{bmatrix} \frown{\boldsymbol{X}} \\ \frown{\boldsymbol{Z}} \end{bmatrix} = \begin{bmatrix} \boldsymbol{f}(\boldsymbol{X}'_o, \boldsymbol{\theta}_o) + \boldsymbol{J}_{X'}(\boldsymbol{X}' - \boldsymbol{X}'_o) + \boldsymbol{J}_\theta\left(\hat{\boldsymbol{\theta}} - \boldsymbol{\theta}_o\right) \\ \boldsymbol{f}(\boldsymbol{Z}'_o, \boldsymbol{\theta}_o) + \boldsymbol{J}_{Z'}(\boldsymbol{Z}' - \boldsymbol{Z}'_o) + \tilde{\boldsymbol{J}}_\theta\left(\hat{\boldsymbol{\theta}} - \boldsymbol{\theta}_o\right) \end{bmatrix} \tag{25}$$

which, to a first-order approximation, mimic the result of the traditional stepwise approach, that is,

$$\begin{bmatrix} \widehat{X} \\ \widehat{Z} \end{bmatrix} \simeq \begin{bmatrix} f\left(X', \hat{\theta}\right) \\ f\left(Z', \hat{\theta}\right) \end{bmatrix} \tag{26}$$

All previous estimators refer to a single execution of the weighted LSA in the linearized system of Eq. (18). Their use in practical applications with nonlinear transformation models requires a recursive algorithm, as explained in more detail in Section 4.2.

4.1.3 Basic features of the stacking approach

Compared to the traditional stepwise methodology, the stacking approach leads to the same least squares estimate for the transformation parameters but to different values for the estimated coordinates in the target frame. This partial equivalency is expected since the inclusion of NPs into the adjustment procedure does not contribute additional information for the transformation parameters. On the other hand, the estimated coordinates contain extra corrections which are derived from stochastic filtering of the coordinate residuals $X - \widehat{X}$ and kriging-like prediction over all points of interest [see Eq. (24)]. Loosely speaking, the effect of those corrections resembles a rubber-sheeting process in the sense of "stretching" the classic stepwise solution to counteract the propagated data noise in the entire set of transformed coordinates.

The stacking approach permits also the exact fit over all CPs regardless of the noise level in the initial frame. This essential property is easily verified by Eq. (24) which implies that

$$\boldsymbol{\Sigma}_X = \mathbf{0} \quad \rightarrow \quad \hat{X} = X \tag{27}$$

or in a loosened version

$$\boldsymbol{\Sigma}_X \ll J_{X'} \boldsymbol{\Sigma}_{X'} J_{X'}^T \quad \rightarrow \quad \hat{X} \simeq X \tag{28}$$

The first condition dictates that the transformed coordinates of CPs will match their prior values, if the latter are assumed to be of perfect quality. The second condition is also useful for practical applications, as it allows the users to improve the fitting performance of the transformation results via a simple tuning of the CV matrix $\boldsymbol{\Sigma}_X$. This last option is essentially equivalent to *stochastic constraining* of the prior coordinates of CPs in the target frame.

As a final note, let us point out that both approaches give similar results in the presence of noiseless data in the initial frame. In such case the least squares estimators of the previous section admit the conditional behavior:

$$\boldsymbol{\Sigma}_{X'} = \mathbf{0} \quad \rightarrow \quad \hat{\boldsymbol{v}}_{X'} = \mathbf{0} \quad \rightarrow \quad \hat{X} = \widehat{X} \simeq f\left(X', \hat{\theta}\right) \tag{29}$$

$$\boldsymbol{\Sigma}_{Z'X'} = \mathbf{0} \quad \rightarrow \quad \hat{\boldsymbol{v}}_{Z'} = \mathbf{0} \quad \rightarrow \quad \hat{Z} = \widehat{Z} \simeq f\left(Z', \hat{\theta}\right). \tag{30}$$

Interestingly, the CV matrix $\boldsymbol{\Sigma}_{Z'}$ does not play an active role within the stacking approach, in contrast to the cross-CV matrix $\boldsymbol{\Sigma}_{Z'X'}$ which is of crucial importance for the optimal transformation at the NPs [see Eq. (22)]. In **Table 1** all relevant cases that can appear in SCT problems are classified with regard to the stochastic model of the observed coordinates in the respective frames.

	CV matrices of observed coordinates		Does data noise filtering occur in the transformation process ?
	Initial frame	Target frame	
Control points	$\Sigma_{X'} \neq 0$	$\Sigma_X \neq 0$	Yes
	$\Sigma_{X'} \neq 0$	$\Sigma_X = 0$	Yes—perfect fit to prior values
	$\Sigma_{X'} = 0$	$\Sigma_X \neq 0$	No
Non-control points	$\Sigma_{Z'} \neq 0$ $\Sigma_{Z'X'} \neq 0$		Yes
	$\Sigma_{Z'} \neq 0$ $\Sigma_{Z'X'} = 0$		No
	$\Sigma_{Z'} = 0$ $\Sigma_{Z'X'} = 0$		No

Table 1.
Different cases in the stacking approach with regard to the stochastic model of the observed coordinates.

4.2 Computational aspects

The numerical computation of the stacking solution in nonlinear SCT problems requires a recursive implementation of the least squares estimators given in Section 4.1.2. The Newton-Gauss iteration method is suitable for this purpose and entails the updating of the approximate vectors X'_o, Z'_o, θ_o at each step by their adjusted values from the previous step until sufficient convergence is achieved in all estimated quantities of interest [3, 24, 34].

The aforesaid procedure should be applied for computing both the transformation parameters and the coordinates of CPs/NPs in the target frame, based on the following algorithm:

$$\hat{\theta}^{(k)} = \theta_o^{(k)} + \left(J_\theta^T W J_\theta\right)^{-1} J_\theta^T W \left(X - f\left(X'^{(k)}_o, \theta_o^{(k)}\right) - J_{X'}\left(X' - X'^{(k)}_o\right)\right) \tag{31}$$

$$\begin{bmatrix} \widehat{X}^{(k)} \\ \widehat{Z}^{(k)} \end{bmatrix} = \begin{bmatrix} f\left(X'^{(k)}_o, \theta_o^{(k)}\right) + J_{X'}\left(X' - X'^{(k)}_o\right) + J_\theta\left(\hat{\theta}^{(k)} - \theta_o^{(k)}\right) \\ f\left(Z'^{(k)}_o, \theta_o^{(k)}\right) + J_{Z'}\left(Z' - Z'^{(k)}_o\right) + \tilde{J}_\theta\left(\hat{\theta}^{(k)} - \theta_o^{(k)}\right) \end{bmatrix} \tag{32}$$

$$\begin{bmatrix} \hat{v}_{X'}^{(k)} \\ \hat{v}_{Z'}^{(k)} \end{bmatrix} = \begin{bmatrix} \Sigma_{X'} J_{X'}^T \\ \Sigma_{Z'X'} J_{X'}^T \end{bmatrix} \left(\Sigma_X + J_{X'} \Sigma_{X'} J_{X'}^T\right)^{-1} \left(X - \widehat{X}^{(k)}\right) \tag{33}$$

$$\begin{bmatrix} \hat{X}^{(k)} \\ \hat{Z}^{(k)} \end{bmatrix} = \begin{bmatrix} \widehat{X}^{(k)} \\ \widehat{Z}^{(k)} \end{bmatrix} + \begin{bmatrix} J_{X'} \hat{v}_{X'}^{(k)} \\ J_{Z'} \hat{v}_{Z'}^{(k)} \end{bmatrix} \tag{34}$$

where the index $k = 1, 2, \ldots$ denotes the LSA iteration step. All Jacobian matrices shown in these equations should be re-evaluated at each step as follows:

$$J_{X'} = \frac{\partial f(X', \theta)}{\partial X'}\bigg|_{\theta_o^{(k)},\, X'^{(k)}_o}, \quad J_{Z'} = \frac{\partial f(Z', \theta)}{\partial Z'}\bigg|_{\theta_o^{(k)},\, Z'^{(k)}_o} \tag{35}$$

$$J_\theta = \frac{\partial f(X', \theta)}{\partial \theta}\bigg|_{\theta_o^{(k)},\, X'^{(k)}_o}, \quad \tilde{J}_\theta = \frac{\partial f(Z', \theta)}{\partial \theta}\bigg|_{\theta_o^{(k)},\, Z'^{(k)}_o}. \tag{36}$$

Note that the auxiliary weight matrix $\boldsymbol{W}$ that appears in Eq. (31) depends on $\boldsymbol{J}_{X'}$ [see Eq. (5)] and it is also required to be updated at each step.

To initialize the Newton-Gauss iteration process, a simple choice is to set the approximate coordinates equal to the observed values ($\boldsymbol{X}'_o = \boldsymbol{X}'$, $\boldsymbol{Z}'_o = \boldsymbol{Z}'$), while the approximate transformation parameters are typically obtained via empirical procedures. The initial computation of $\hat{\boldsymbol{\theta}}$ is thus reduced to the simpler form given already in Section 3.2, whereas for subsequent iterations the rigorous expression of Eq. (31) should be used. The updating of all approximate vectors at each step should be performed by the following equations:

$$\boldsymbol{\theta}_o^{(k)} = \hat{\boldsymbol{\theta}}^{(k-1)} \tag{37}$$

$$\boldsymbol{X}'^{(k)}_o = \boldsymbol{X}' + \hat{\boldsymbol{v}}_{X'}^{(k-1)} \tag{38}$$

$$\boldsymbol{Z}'^{(k)}_o = \boldsymbol{Z}' + \hat{\boldsymbol{v}}_{Z'}^{(k-1)} \tag{39}$$

Special cases with noise-free coordinates in the initial frame ($\boldsymbol{\Sigma}_{X'} = \boldsymbol{0}$) and/or uncorrelated coordinates between CPs and NPs ($\boldsymbol{\Sigma}_{Z'X'} = \boldsymbol{0}$) can be easily treated under the previous framework, and they lead to identical results as the traditional stepwise approach.

4.3 Statistical accuracy assessment in SCT solutions

The error CV matrices of $\hat{\boldsymbol{\theta}}$, $\hat{\boldsymbol{X}}$, and $\hat{\boldsymbol{Z}}$ are the fundamental elements for the formal quality assessment in SCT solutions. Their rigorous expressions are obtained by covariance propagation to the respective estimators given in previous sections, and they are presented here without their full mathematical proofs.

Both the stepwise and the stacking approach lead to the same optimal estimate for the transformation parameters, whose error CV matrix is given by the formula:

$$\boldsymbol{\Sigma}_{\hat{\theta}} = \left(\boldsymbol{J}_\theta^T \left(\boldsymbol{\Sigma}_X + \boldsymbol{J}_{X'}\boldsymbol{\Sigma}_{X'}\boldsymbol{J}_{X'}^T\right)^{-1}\boldsymbol{J}_\theta\right)^{-1}. \tag{40}$$

Regarding the accuracy assessment of the transformed coordinates by the stepwise approach, the following expressions should be used:

$$\begin{aligned}\overline{\boldsymbol{\Sigma}}_{\hat{X}} = {} & \boldsymbol{J}_\theta\boldsymbol{\Sigma}_{\hat{\theta}}\boldsymbol{J}_\theta^T + \boldsymbol{J}_{X'}\boldsymbol{\Sigma}_{X'}\boldsymbol{J}_{X'}^T - \left(\boldsymbol{J}_\theta\boldsymbol{\Sigma}_{\hat{\theta}}\boldsymbol{J}_\theta^T\right)\left(\boldsymbol{\Sigma}_X + \boldsymbol{J}_{X'}\boldsymbol{\Sigma}_{X'}\boldsymbol{J}_{X'}^T\right)^{-1}\left(\boldsymbol{J}_{X'}\boldsymbol{\Sigma}_{X'}\boldsymbol{J}_{X'}^T\right) \\ & -\left(\boldsymbol{J}_{X'}\boldsymbol{\Sigma}_{X'}\boldsymbol{J}_{X'}^T\right)\left(\boldsymbol{\Sigma}_X + \boldsymbol{J}_{X'}\boldsymbol{\Sigma}_{X'}\boldsymbol{J}_{X'}^T\right)^{-1}\left(\boldsymbol{J}_\theta\boldsymbol{\Sigma}_{\hat{\theta}}\boldsymbol{J}_\theta^T\right)\end{aligned} \tag{41}$$

$$\begin{aligned}\overline{\boldsymbol{\Sigma}}_{\hat{Z}} = {} & \tilde{\boldsymbol{J}}_\theta\boldsymbol{\Sigma}_{\hat{\theta}}\tilde{\boldsymbol{J}}_\theta^T + \boldsymbol{J}_{Z'}\boldsymbol{\Sigma}_{Z'}\boldsymbol{J}_{Z'}^T - \left(\tilde{\boldsymbol{J}}_\theta\boldsymbol{\Sigma}_{\hat{\theta}}\boldsymbol{J}_\theta^T\right)\left(\boldsymbol{\Sigma}_X + \boldsymbol{J}_{X'}\boldsymbol{\Sigma}_{X'}\boldsymbol{J}_{X'}^T\right)^{-1}\left(\boldsymbol{J}_{X'}\boldsymbol{\Sigma}_{X'Z'}\boldsymbol{J}_{Z'}^T\right) \\ & -\left(\boldsymbol{J}_{Z'}\boldsymbol{\Sigma}_{Z'X'}\boldsymbol{J}_{X'}^T\right)\left(\boldsymbol{\Sigma}_X + \boldsymbol{J}_{X'}\boldsymbol{\Sigma}_{X'}\boldsymbol{J}_{X'}^T\right)^{-1}\left(\boldsymbol{J}_\theta\boldsymbol{\Sigma}_{\hat{\theta}}\tilde{\boldsymbol{J}}_\theta^T\right)\end{aligned} \tag{42}$$

which refer to the CPs and NPs, respectively. The overbar symbol is used to distinguish the above error CV matrices from the respective expressions that apply in the stacking approach. The latter are given by the general formulae:

$$\boldsymbol{\Sigma}_{\hat{X}} = \overline{\boldsymbol{\Sigma}}_{\hat{X}} - \boldsymbol{K}\boldsymbol{\Sigma}_e\boldsymbol{K}^T \tag{43}$$

$$\boldsymbol{\Sigma}_{\hat{Z}} = \overline{\boldsymbol{\Sigma}}_{\hat{Z}} - \boldsymbol{Q}\boldsymbol{\Sigma}_e\boldsymbol{Q}^T \tag{44}$$

where the auxiliary matrices $\boldsymbol{K}$ and $\boldsymbol{Q}$ are defined as

$$\boldsymbol{K} = \left(\boldsymbol{J}_{X'}\boldsymbol{\Sigma}_{X'}\boldsymbol{J}_{X'}^T\right)\left(\boldsymbol{\Sigma}_X + \boldsymbol{J}_{X'}\boldsymbol{\Sigma}_{X'}\boldsymbol{J}_{X'}^T\right)^{-1} \tag{45}$$

$$\boldsymbol{Q} = \left(\boldsymbol{J}_{Z'}\boldsymbol{\Sigma}_{Z'X'}\boldsymbol{J}_{X'}^T\right)\left(\boldsymbol{\Sigma}_X + \boldsymbol{J}_{X'}\boldsymbol{\Sigma}_{X'}\boldsymbol{J}_{X'}^T\right)^{-1} \tag{46}$$

and $\boldsymbol{\Sigma}_e$ is the CV matrix of the coordinate residuals $\boldsymbol{X} - f\left(\boldsymbol{X}', \hat{\boldsymbol{\theta}}\right)$, that is,

$$\boldsymbol{\Sigma}_e = \boldsymbol{\Sigma}_X + \boldsymbol{J}_{X'}\boldsymbol{\Sigma}_{X'}\boldsymbol{J}_{X'}^T - \boldsymbol{J}_\theta\boldsymbol{\Sigma}_{\hat{\theta}}\boldsymbol{J}_\theta^T. \tag{47}$$

Equations (43) and (44) reveal the expected improvement of the statistical accuracy in the SCT solution by the stacking approach. The diagonal elements (i.e., coordinate error variances) of $\boldsymbol{\Sigma}_{\hat{X}}$ and $\boldsymbol{\Sigma}_{\hat{Z}}$ are always smaller than the respective elements of $\overline{\boldsymbol{\Sigma}}_{\hat{X}}$ and $\overline{\boldsymbol{\Sigma}}_{\hat{Z}}$, a fact that is attributed to the noise filtering of the observed coordinates during the transformation process.

5. Numerical example

To demonstrate the potential of the stacking approach in practical transformation problems, a simple example is given here for a simulated 2D network with seven CPs and four NPs. The true coordinates of all network points are listed in **Table 2**, and they are related by a second-order polynomial transformation:

$$x_i = a_o + a_1 x_i' + a_2 y_i' + a_3 x_i' y_i' + a_4 x_i'^2 + a_5 y_i'^2 \tag{48}$$

$$y_i = b_o + b_1 x_i' + b_2 y_i' + b_3 x_i' y_i' + b_4 x_i'^2 + b_5 y_i'^2 \tag{49}$$

whose associated parameters are provided in **Table 3**.

The observed coordinates for our experiments stem by adding simulated Gaussian noise to the true values of **Table 2**. The known coordinates of NPs in

	Initial frame		Target frame	
	x' (m)	y' (m)	x (m)	y (m)
CP1	100.000	250.000	146.000	287.000
CP2	200.000	423.205	210.768	467.597
CP3	286.602	373.205	239.979	435.802
CP4	157.735	150.000	181.119	177.203
CP5	125.000	200.000	159.250	231.438
CP6	225.000	250.000	222.875	294.188
CP7	250.000	400.000	226.250	452.750
NP1	200.000	300.000	209.250	342.500
NP2	159.000	230.000	184.574	264.665
NP3	220.000	340.000	217.850	386.660
NP4	170.000	270.000	192.750	308.030

Table 2.
True coordinates of CPs and NPs in the simulated test network with respect to the initial and target frame.

the target frame are not included in the observables, but they were used only for cross-validation of the transformation results. The generated random errors at the CPs *in the target frame* are uncorrelated with a common standard deviation of 0.1 cm for the x and y coordinates. On the other hand, the generated random errors at the CPs/NPs *in the initial frame* are spatially correlated in terms of the simplified Gaussian-type covariance model:

$$\sigma_{x_i'x_k'} = \sigma_{y_i'y_k'} = \sigma^2 e^{-A\left(x_i'-x_k'\right)^2 - B\left(y_i'-y_k'\right)^2} \quad (50)$$

$$\sigma_{x_i'y_k'} = \rho\, \sigma^2 e^{-A\left(x_i'-x_k'\right)^2 - B\left(y_i'-y_k'\right)^2} \quad (51)$$

a_o	a_1	a_2	a_3	a_4	a_5
10.25	1.20	0.20	−0.0013	−0.0008	0.0001
b_o	b_1	b_2	b_3	b_4	b_5
18.50	−0.25	1.20	−0.0002	0.0011	−0.0002

Table 3.
True parameter values of the second-order polynomial transformation model.

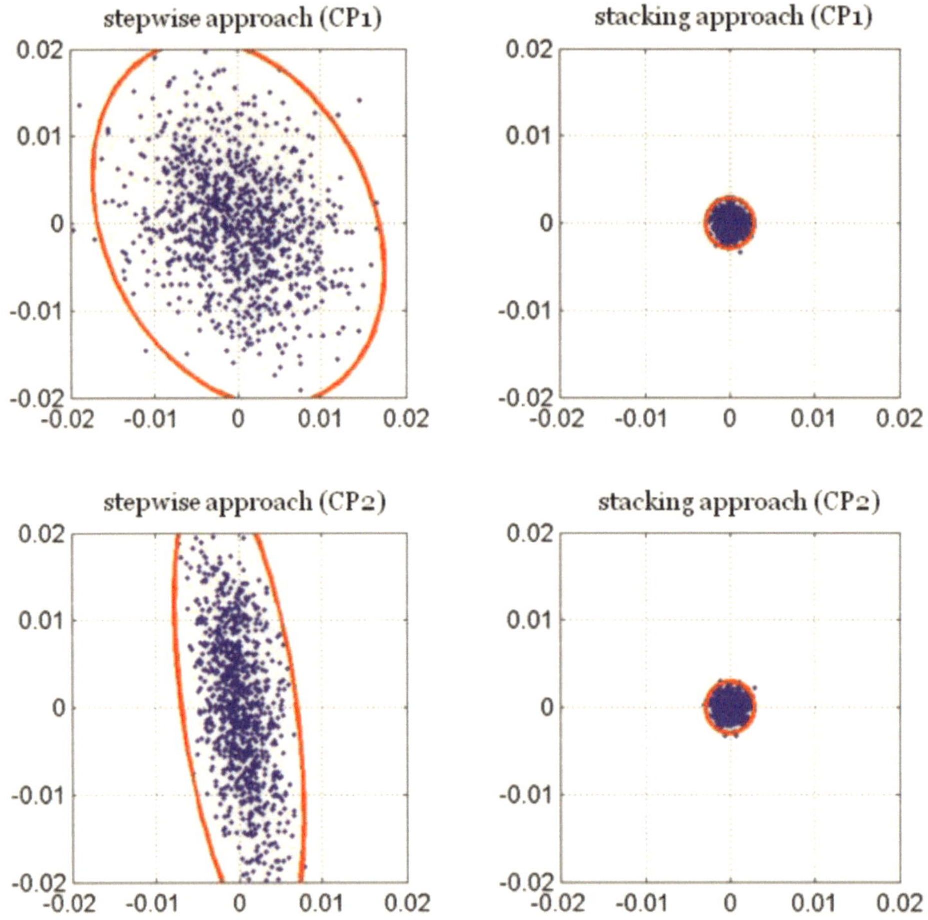

Figure 1.
Differences between the true and the transformed coordinates at two CPs over 1000 Monte Carlo sampling experiments in the simulated network. The point error ellipses (99% confidence level) by each transformation approach are also shown in red color. The scaling of the horizontal axes is in meters.

where σ is the common error standard deviation for the x' and y' coordinates (set equal to 5 cm) and ρ is their error correlation coefficient at each point (set equal to -0.2). The values of the auxiliary parameters A and B were fixed to 6×10^{-7} and 7×10^{-6}, respectively, which ensure the positive definiteness of the resulting CV matrix for the observed coordinates in the initial frame.

Using a Monte Carlo sampling scheme and a Cholesky-based algorithm for the stochastic simulation of correlated random vectors, a total of 1000 noisy ensembles were produced for the triplet of coordinate vectors $\boldsymbol{X}$, $\boldsymbol{X}'$, and $\boldsymbol{Z}'$. These synthetic datasets were used with the stepwise and stacking approach to determine the transformed coordinates and their associated accuracy, over all points of the simulated network.

The differences between the true and the transformed coordinates in the target frame, as obtained by all data ensembles under each approach, are shown in **Figures 1** and **2**. The cloud plots in these figures refer only to a subset of the CPs/NPs, yet similar results are acquired at all other network points. It is clear that the stacking approach yields significantly better results than the traditional stepwise approach, and it effectively filters the existing noise of the initial coordinates. The accuracy improvement ranges from 88 to 92% at the CPs, while it is a bit lower (63–78%) at the NPs (see detailed results in **Table 4**).

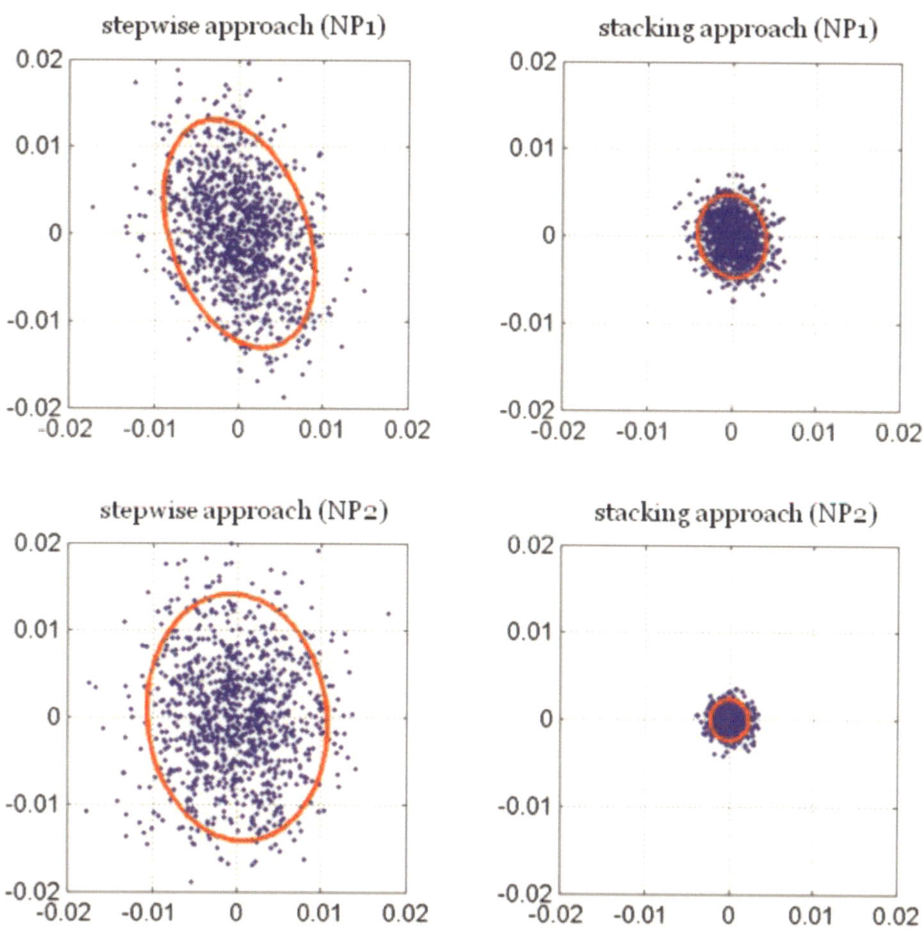

Figure 2.
Differences between the true and the transformed coordinates at two NPs over 1000 Monte Carlo sampling experiments in the simulated network. The point error ellipses (99% confidence level) by each transformation approach are also shown in red color. The scaling of the horizontal axes is in meters.

	Stepwise approach			Stacking approach			Accuracy improvement (%)
	$\sigma_{\hat{x}}$	$\sigma_{\hat{y}}$	$\sqrt{\sigma_{\hat{x}}^2+\sigma_{\hat{y}}^2}$	$\sigma_{\hat{x}}$	$\sigma_{\hat{y}}$	$\sqrt{\sigma_{\hat{x}}^2+\sigma_{\hat{y}}^2}$	
CP1	0.6	0.7	0.9	0.1	0.1	0.1	89
CP2	0.3	0.9	0.9	0.1	0.1	0.1	89
CP3	0.3	0.7	0.7	0.1	0.1	0.1	86
CP4	0.6	1.0	1.2	0.1	0.1	0.1	92
CP5	0.6	0.8	1.0	0.1	0.1	0.1	90
CP6	0.4	0.6	0.8	0.1	0.1	0.1	88
CP7	0.3	0.8	0.8	0.1	0.1	0.1	88
NP1	0.5	0.7	0.8	0.2	0.2	0.3	63
NP2	0.5	0.7	0.9	0.1	0.1	0.2	78
NP3	0.4	0.7	0.8	0.2	0.3	0.3	63
NP4	0.5	0.7	0.8	0.2	0.2	0.3	63

All values given in cm.

Table 4.
Standard deviations of the transformed coordinates in the simulated network by different approaches.

It should be emphasized that the stochastic model of the observed coordinates plays a key role in the performance of the stacking approach. This means that the results shown here may exhibit different behavior—displaying either insignificant or even more profound accuracy improvement for the transformed coordinates—for varied choices of the CV matrices $\boldsymbol{\Sigma}_X$, $\boldsymbol{\Sigma}_{X'}$, $\boldsymbol{\Sigma}_{Z'}$, and $\boldsymbol{\Sigma}_{X'Z'}$.

Author details

Christopher Kotsakis
Department of Geodesy and Surveying, School of Rural and Surveying Engineering, Faculty of Engineering, Aristotle University of Thessaloniki, Thessaloniki, Greece

*Address all correspondence to: kotsaki@topo.auth.gr

References

[1] Ayhan E, Akar OE, Uzun S, Dilaver A, Kansu O. Analysis of digital data obtained from raster and vector maps. Journal of Surveying Engineering. 2011; **137**:65-69

[2] Beinat A, Crosilla F. Generalised procrustes algorithms for the conformal updating of a cadastral map. Zeitschrift für Geodäsie. 2003;**128**(5):341-349

[3] Bjorck A. Numerical Methods for Least Squares Problems. Philadelphia PA: SIAM; 1996. p. 408

[4] Cobb MA, Chung MJ, Poley H, Petry FE, Shaw KB, Miller HV. A rule-based approach for the conflation of attributed vector data. GeoInformatica. 1998;**2**(1): 7-35

[5] Doytsher Y, Hall JK. Gridded affine transformation and rubber-sheeting algorithm with Fortran program for calibrating scanned hydrographic survey maps. Computers and Geosciences. 1997;**23**(7):785-791

[6] Doytsher Y, Filin S, Ezra E. Transformation of datasets in linear-based map conflation framework. Surveying and Land Information Systems. 2001;**61**(3):159-169

[7] Eggert DW, Lorusso A, Fisher RB. Estimating 3-D rigid body rotations: A comparison of four major algorithms. Machine Vision and Applications. 1997; **9**:272-290

[8] Felus YA. On the positional enhancement of digital cadastral maps. Survey Review. 2007;**39**(306):268-281

[9] Fruh C, Zakhor A. Constructing 3D city models by merging aerial and ground views. IEEE Computer Graphics and Applications. 2003;**23**(6):52-61

[10] Gill PE, Murray W. Algorithms for the solution of the nonlinear least-squares problem. SIAM Journal on Numerical Analysis. 1978;**15**(5):977-992

[11] Grafarend EW, Awange JL. Nonlinear analysis of the three-dimensional datum transformation conformal group $C_7(3)$. Journal of Geodesy. 2003;**77**:66-76

[12] Grafarend EW. Linear and Nonlinear Models—Fixed Effects, Random Effects and Mixed Models. Berlin: de Gruyter; 2006. p. 752

[13] Goryn D, Hein S. On the estimation of rigid body rotation from noisy data. IEEE Transactions on Pattern Analysis and Machine Intelligence. 1995;**17**: 1219-1220

[14] Hackeloeer A, Klasing K, Krisp JM, Meng L. Georeferencing: A review of methods and applications. Annals of GIS. 2014;**20**(1):61-69

[15] Hild H, Fritsch D. Integration of vector data and satellite imagery for geocoding. International Archives of Photogrammetry and Remote Sensing. 1998;**32**(4):246-251

[16] Klebanov M, Doytsher Y. Cadastral triangulation: A block adjustment approach for joining numerous cadastral blocks. Nordic Journal of Surveying and Real Estate Research, Special Series. 2009;**4**:53-68

[17] Koch K-R. Parameter Estimation and Hypothesis Testing in Linear Models. 2nd ed. Berlin Heidelberg: Springer-Verlag; 1999. p. 331

[18] Kotsakis C, Vatalis A, Sanso F. On the importance of intra-frame and inter-frame covariances in frame transformation theory. Journal of Geodesy. 2014;**88**(12):1187-1201

[19] Leick A, van Gelder BHW. On similarity transformations and geodetic

network distortions based on Doppler satellite observations. Technical Report No. 235, Department of Geodetic Science, the Ohio State University, Columbus, Ohio; 1975. 145pp

[20] Lichti DD. Terrestrial laser scanner self-calibration: Correlation sources and their mitigation. ISPRS Journal of Photogrammetry and Remote Sensing. 2010;**65**:93-102

[21] Lu L, Zhang Y, Tao P, Zhang Z, Zhang Y. Estimation of transformation parameters between centre-line vector road maps and high resolution satellite images. The Photogrammetric Record. 2013;**28**(142):130-144

[22] Mitishita E, Cortes J, Centeno J. Indirect georeferencing of digital SLR imagery using signalized Lidar control points. The Photogrammetric Record. 2011;**26**(133):58-72

[23] Molnar G, Pfeifer N, Ressl C, Dorninger P, Nothegger C. Range calibration of terrestrial laser scanners with piecewise linear functions. Photogrammetrie, Fernerkundung, Geoinformation. 2009;**1**:9-21

[24] Pope A. Some pitfalls to be avoided in the iterative adjustment of nonlinear problems. In: Proceedings of the 38th Annual ASPRS Meeting. Falls Church, VA: American Society of Photogrammetry; 1972. pp. 449-477

[25] Rabbani T, Dijkman S, van den Heuvel F, Vosselman G. An integrated approach for modeling and global registration of point clouds. ISPRS Journal of Photogrammetry and Remote Sensing. 2007;**61**:355-370

[26] Reshetyuk Y. A unified approach to self-calibration of terrestrial laser scanners. ISPRS Journal of Photogrammetry and Remote Sensing. 2010;**65**:445-456

[27] Ruiz JJ, Ariza FJ, Urena MA, Blazquez EB. Digital map conflation: A review of the process and a proposal for classification. International Journal of Geographical Information Science. 2011; **25**(9):1439-1466

[28] Sanso F, Venuti G. Inverse cartographic problems: Theory and first results. Bollettino di Geodesia e Scienze Affini Anno LVII. 1998;**4**:437-450

[29] Sequeira V, Ng K, Wolfart E, Goncalves J, Hogg D. Automated reconstruction of 3D models from real environments. ISPRS Journal of Photogrammetry and Remote Sensing. 1999;**55**(1):1-22

[30] Soler T, Snay RA. Transforming positions and velocities between the international terrestrial reference frame of 2000 and North American datum of 1983. Journal of Surveying Engineering. 2004;**130**(2):49-55

[31] Surmann H, Nuchter A, Hertzberg J. An autonomous mobile robot with a 3D laser range finder for 3D exploration and digitalization of indoor environments. Robotics and Autonomous Systems. 2003;**45**(3–4): 181-198

[32] Tamim N, Schaffrin B. A methodology to create a digital cadastral overlay through upgrading digitized cadastral data. Surveying and Land Information Systems. 1995;**55**(1):3-12

[33] Teunissen PJG. Adjusting and testing with the models of the affine and similarity transformations. Manuscripta Geodaetica. 1986;**11**:214-225

[34] Teunissen PJG. Nonlinear least squares. Manuscripta Geodaetica. 1990; **15**(3):137-150

[35] Torres-Martinez JA, Seddaiu M, Rodriguez-Gonzalvez P, Hernandez-Lopez D, Gonzalez-Aguilera D. A multi-data source and multi-sensor approach for the 3D reconstruction and web visualization of a complex archaeological site: The case study of

"Tolmo de Minateda". Remote Sensing. 2016;**8**(7):550. DOI: 10.3390/rs8070550

[36] Umeyama S. Least-squares estimation of transformation parameters between two point patterns. IEEE Transactions on Pattern Analysis and Machine Intelligence. 1991;**13**(4): 376-380

[37] Watson GA. Computing Helmert transformations. Journal of Computational and Applied Mathematics. 2006;**197**(2):387-394

[38] White MS, Griffin P. Piecewise linear rubber-sheet map transformation. The American Cartographer. 1985;**12**(2): 123-131

[39] Zitova B, Flusser J. Image registration methods: A survey. Image and Vision Computing. 2003;**21**: 977-1000